No Surrender

MISSISSIPPI KU-KLUX IN THE DISGUISES IN WHICH THEY WERE CAPTURED.
[FROM A PHOTOGRAPH.]

The Mississippi Ku-Klux in the disguises in which they were captured. (Library of Congress)

No Surrender

ASYMMETRIC WARFARE IN THE RECONSTRUCTION SOUTH, 1868–1877

Keith D. Dickson

 PRAEGER™

An Imprint of ABC-CLIO, LLC
Santa Barbara, California • Denver, Colorado

Library of Congress Cataloging-in-Publication Data

Names: Dickson, Keith D., author.
Title: No surrender : asymmetric warfare in the Reconstruction South, 1868–1877 /
   Keith D. Dickson.
Description: Santa Barbara, California : Praeger, 2017. | Includes bibliographical
   references and index. | Description based on print version record and CIP data
   provided by publisher; resource not viewed.
Identifiers: LCCN 2017002093 (print) | LCCN 2017017126 (ebook) |
   ISBN 9781440848940 (ebook) | ISBN 9781440848933 (hard copy : alk. paper)
Subjects: LCSH: Reconstruction (U.S. history, 1865–1877) | Asymmetric warfare—
   Southern States—History—19th century.
Classification: LCC E668 (ebook) | LCC E668 .D549 2017 (print) | DDC 973.8—dc23
LC record available at https://lccn.loc.gov/2017002093

ISBN: 978-1-4408-4893-3
EISBN: 978-1-4408-4894-0

21 20 19 18 17   1 2 3 4 5

This book is also available as an eBook.

Praeger
An Imprint of ABC-CLIO, LLC

ABC-CLIO, LLC
130 Cremona Drive, P.O. Box 1911
Santa Barbara, California 93116-1911
www.abc-clio.com

This book is printed on acid-free paper ∞

Manufactured in the United States of America

To Karen:
The Happiness, Hope, and Love of My Life

Contents

Acknowledgments

I would like to thank Jim Ciment for his assistance and encouragement in developing the idea for this book. I am grateful to the members of the editorial board at Praeger for their interest in this project and to the members of the editorial staff for their careful and thoughtful assistance. I am indebted to my colleagues in the History Department at the Joint Advanced Warfighting School for their support and encouragement. I am particularly grateful to Mike Pavelec for his indomitable spirit and unbounded enthusiasm for my efforts and to Bryon Greenwald for his always reasonable and thoughtful insights. Gregory Miller added his own unique point of view, never allowing me to take myself too seriously. Together, this group spent many pleasurable hours in deep discussion of innumerable topics, some even relevant to this book. Each of my colleagues has served as a personal inspiration and a model of teaching and scholarship.

This has been an undertaking accomplished over many very late nights and early mornings, with work periods usually jammed in between teaching, course preparation, and grading. Weekends have been a series of closed off compartments of time where everything else took second and third priority. The process of research and writing drives the author away from everything else not related to the project in the oftentimes all-consuming effort to capture ideas and express them adequately. My wife Karen, as is her wont, played no role in this effort. As always, she keeps her distance and waits out the storms, knowing that there is an end and that the sun will shine again. So to her I dedicate this book, recognizing full well that I must now implement my own domestic reconstruction strategy.

Introduction: Reconstruction Reconsidered in a New Light

It has often been said that the American Civil War is the first modern war or, at least, the harbinger of the devastating total wars of the twentieth century. Many historians cite the employment of new technologies that changed the conduct of war: from railroads and the telegraph, which expanded war to a new level from the pure tactical to the operational. New weapons of war, from armored ships to the breech-loading rifle, changed the methods of warfare, as well as created new tactics and new requirements for commanders. In the same manner, it can also be said that Reconstruction was another harbinger of modern war: presaging the increasingly difficult challenge to forge peace and stability from military victory. These challenges manifested themselves in a number of very familiar ways: the uncertain transition from military oversight to civilian oversight; the multidimensional character of resistance; and the frustrations of power, and its inability to achieve desired goals.

The Southern novelist David Madden has asserted that Americans are missing the broader meaning of the Civil War and Reconstruction because we who have spent our lives studying it have failed, as he puts it, "to untangle the web of facts we know and view them in the richest possible context."[1] Madden offered a challenge to every writer in every discipline to take up a number of questions for understanding the Civil War and Reconstruction in the twenty-first century. A variation of one of those questions is the reason for this book: How does Reconstruction in the South, Madden asked, compare with our reconstruction efforts in enemy countries? Although he was referring to reconstruction in Europe after the World War II, he did not mention U.S. nation-building efforts in twenty-first century wars, particularly Afghanistan and Iraq. It is the reconstruction effort in the aftermath of Operation Iraqi Freedom in 2003 that is most closely connected to reconstruction in the American South. Indeed, it was during my service as the combat historian for U.S. Special Operations Command in Iraq in 2003 that I had the sense of stepping back into history. It struck me clearly and plainly: what was happening before my eyes was what had happened in the South. Just as federal officers must have looked at the violence and political chaos in Louisiana, South Carolina, and Georgia and wondered what had happened to victory, I had the same sense of confusion. As I found myself wondering what we had missed that created such rapid and drastic change, I immediately sensed a bond that stretched across 150 years of time.

As a result of the confusion and frustration with what the coalition forces faced, the U.S. government and the U.S. military began to throw words at what was happening in Iraq between 2003 and 2007. It was an insurgency. It was unconventional war. It was terrorist-based violence. It was irregular conflict. What was opaque to the leadership, who inexplicably decided to reapply French counterinsurgency techniques from Algeria in the 1960s to the situation in Iraq, was that this phenomenon the coalition had encountered was a continuum from *war* to *warfare*. What it needed was a name and a structured means of understanding its dimensions, its purpose, and its components. As I looked at the past to help understand the present, I saw parallel after parallel with Reconstruction in the South from 1865 to 1877. Two terms that had some currency in the 1990s for a short time within the defense community—asymmetry and asymmetric warfare—gave me what I needed. I could now approach the tangled web and put the strands in a contemporary context to help explain why and how Reconstruction in the South unfolded as it did and provide new perspectives and appreciation for the importance of the Civil War and Reconstruction in our national life today.

Untangling the web of facts about Reconstruction is a challenge, especially when viewed in terms of the wide number of approaches historians have taken in the twentieth century. Rather than untangling this web of facts, the web has ensnared historians, leading them to slash and cut in frustration, and, from the tatters, assemble an entirely new web of interpretation heavily infused with moral judgment. The result of this reassembly is a descriptive narrative of events in which the historian names and arranges the historical figures (presidents, congressmen, radicals, scalawags, carpetbaggers, Negroes, freedpeople, Democrats, Planter aristocracy, conservatives, Ku-Klux, racists, white supremacists, terrorists, guerrillas, etc.) like dolls on a miniature stage. The historian then selectively moves the dolls to give one or the other a certain amount of time in the spotlight. But because the interplay of the figures is so immensely complicated, the historian must decide how much time one player, or group of players, remains in the spotlight. Thus, the moral of the story—the explanation for what happened—is related to those players with the most time in the spotlight.

This process is easy to trace in a brief overview of the historiography of Reconstruction. The Dunning School, named after the Columbia University historian William Archibald Dunning (1857–1922), was composed of Dunning's students who, later as historians themselves, produced the first studies on Reconstruction at the state level. They placed incompetent Negroes, opportunist scalawags, and corrupt carpetbaggers in the spotlight and concluded that Reconstruction was the result of misguided national policies that were an unmitigated tragedy of governmental overreach imposed on prostrate Southerners. The Beardian interpretation, outlined in Charles and Mary Beard's landmark history *The Rise of American Civilization* (1927), was that Reconstruction was the second American Revolution, in which Northern capitalists and congressmen, backed by the power of the national state, took the spotlight and established a favorable direction for the future development of the modern United States. The revisionists, influenced by 1960s social change, also possessed a deep faith in the power of the national government to achieve noble goals, placed activist freedmen, reformist Southern whites, and idealist, progressive Northerners in the spotlight. This alliance, in their view, created state governments that were pregnant with possibilities, but their advancements were only temporary because of blind white racist hatred that could have been easily controlled and stopped with a forceful federal intervention backed by a significant increase in military power deployed to the South.[2] The judgment was that Reconstruction was indeed a tragedy, but only as a golden promise unfulfilled. Post-Revisionists assembled the figures, moved them around, and found that at the end of the story, the Old South had essentially reemerged largely unchanged from the previous period of turmoil and violence. Those who had the power and influence before the war emerged largely intact after Reconstruction. The moral of the story was a sense of regret over lost opportunities—a stillborn revolution.[3]

In a 1982 article, historian Eric Foner presented his opinion that "historians have failed to produce a coherent modern portrait of Reconstruction." Part of the reason for this, he believed, was that American society had failed "to come to terms with the results of the Civil War and the consequences of emancipation."[4] This thesis of blacks redefining their place in American society to allow the country to live up to its essential ideas of rights and freedom became the basis of his now famous and influential book, *Reconstruction: America's Unfinished Revolution, 1863–1877.* Now the dominant and accepted view of the period, it reflects current American sensibilities, which are bound in a comforting and soothing narrative of a spotlight on black agency and autonomous empowerment, and of a time of promise, although stunted, whose seeds of civil and political rights were planted for a future revival.[5]

The Reconstruction period has offered historians such as Foner the opportunity to stand triumphantly on the moral high ground and to present on stage the uncomplicated high promise of freedom, which only needed to be bestowed on the freedpeople to create a new and more perfect and enlightened United States extricated from racial inequalities and set on the path to progress. But this bright moment of promise was curbed and ultimately denied by white Southern hatred. Foner's story has been music to America's ears for nearly a generation. But times have changed, and there is now a need for that "coherent modern portrait" to emerge. Madden's web of facts remains to be arranged now in a different way to provide a richer context in the wake of the end of the Cold War and the outcomes of U.S. power employed in the aftermath of the overwhelming success of military force to reconstruct unfamiliar societies. More than ever, understanding Reconstruction after our great Civil War will help us understand ourselves in today's world.[6]

The current popularly accepted narrative of Reconstruction attacks what is perceived as the dangerous myth that recalls a tragic era overseen by a triumvirate of opportunist, cynical carpetbaggers; vindictive, unscrupulous scalawags; and debased, ignorant blacks that represented a repudiation of national ideals and required white Southerners to redeem their states from the control of this unholy trinity. In its place is perhaps an equally dangerous but more winsome myth of black liberation equality forged by black activists and their noble white allies to create an interracial democracy that was ruined by white racist violence.[7] This narrative must rely on the conveniently all-pervasive paradigm of racism to explain the violent resistance of the white South, reducing the complexities of the postwar South to its most simplistic (and more palatable) formula of victim and perpetrator.

Within this overarching narrative, revisionist historians have assessed the records of the individual states under the Republican governments, and their studies have led them to make the argument that the traditional criticism of the Reconstruction governments was mostly unfounded. They cite the numerous progressive and democratic reform measures in the state constitutions and the dignity and orderly cooperation of the biracial constitutional conventions, as well as the expansion of government social services and schools under largely competent and capable governors. Although they accept that the Reconstruction governments were at times plagued by mismanagement, corruption, and inexperience that often led to ruinous and prodigal expenditures, these are excused as part and parcel of a broader culture of political corruption extending into the Northern municipal city governments, Congress, and the Grant administration. The corresponding enormous public debt and crippling taxes are also often excused as a necessary part of implementing the progressive transformative agenda. The argument is that the constitutions that established these Congressionally mandated Reconstruction governments were not revolutionary, but they represented improvements and essential needed reforms to the antebellum state constitutions.

These assessments, however, are made in a political and social vacuum. They were not just governments functioning during a particular political party's control, whose record can be judged on the balance of achievements compared to shortcomings. Indeed, these were governments established—however orderly, high minded, and moderate in reform and intent—under extraordinary circumstances that were imposed from a central government and instituted under military supervision. These state governments were expected to initiate a social and political transformation that had not occurred anywhere else in the United States and, in fact, was strongly opposed in other parts of the country. These governments were administering a postwar society, devastated by human and material losses and confronting a completely new and disconcerting environment based on the social and political equality of the newly created freedman with the white man. Societies undergoing a revolutionary transition like this should be expected to have high levels of violence, and white Southerners were already conditioned to extreme violence. Therefore, it was not surprising that violence in any number of forms would break out in response to the social and political conditions that had been imposed on white Southerners by a government they saw as illegitimate.[8] More importantly, white Southerners who believed that a political equilibrium had been established under Presidential Reconstruction were suddenly placed in a condition of subordination and isolation under the Congressional Reconstruction Act and its supplements, and they were expected to accept the newly imposed status quo passively.

Moreover, these new state governments were expected to replace entirely existing state governments that had been functioning but were declared provisional by Congressional fiat. At the same time, these new governments were to establish a completely functioning administrative and legal apparatus from a limited pool of politically reliable candidates and were expected to exercise these functions immediately. In a benign environment, these governments would have been challenged enough, but the volatile environment that existed when these governments were established created unexpected and unanticipated challenges to their ability to provide a level of security and order essential to their claims to legitimacy.

When examined in conditions of a political vacuum, the popular revisionist argument that the freedman was the agent of his own freedom is not altogether supportable when viewed from the perspective of asymmetric warfare. Undoubtedly, the freedmen's political actions during Reconstruction were not altogether passive or irresponsible, but this point is irrelevant in the context of asymmetric warfare. The postwar environment demanded from the freedmen a nakedly realist approach of self-interest that led them to support the Republican Party. Votes for the Republicans would most likely directly benefit them, while the act of voting itself was symbolic of constitutionally granted immunities and privileges of citizenship. Republicans in the Union Leagues had prepared the freedmen in following this approach by linking freedom and rights with the Republican Party and emphasizing the importance of the ballot as the means to guarantee their future as citizens as well as their protection. The freedmen were essential to the survival of the Republican state governments, and therefore black voters had to be molded into a reliable cohort to ensure unchallenged Republican control of the state governments in the South for a sufficient length of time to enact the Reconstruction agenda of social, political, and economic reform. In the context of asymmetric warfare, the freedmen were key political actors essential to the survival of the state governments and, therefore, a legitimate target of attack. It was a simple calculus: the fewer Republican votes from freedmen, the less secure the Republican governments became; likewise, the more violence and intimidation directed against the freedmen and their white allies, the weaker and less legitimate the Republican governments appeared.

In asymmetric warfare, there is a battle for a critical threshold of support. The winner dominates the psychological and political landscape. The weaker actor attempts to overcome the dominant actor's advantage of popular support by delegitimizing the government and dissuading or deterring the stronger actor's supporters from sustaining their interest and willingness to continue their support. This is accomplished in any number of ways: threats and intimidation, assassination, highly visible attacks on prominent leaders, attacks on symbols of power, and even random violence to spread fear and insecurity—all supported by an information campaign that seeks to supersede the dominant actor's image of order, control, and legitimacy.

Whether completely true or not, the indelible image of Republican governments, whose record of greed, exploitation, and waste was a disaster for the South, served as a powerful and essential weapon of asymmetric warfare. Southern newspapers gained and maintained dominance of the information battlefield almost immediately, and they not only built strong levels of solidarity and mobilization of the Southern-Confederate collective identity but also gained sympathy in the North and weakened both internal and external support for the Republican state governments. As a result, blacks and loyal whites were less willing to cross the threshold of support because doing so exposed them to retaliatory violence. In contrast, highly visible opposition and defiance to the state and federal authorities, employing both violent and nonviolent political agitation drew many white Southerners over the threshold of support to create a highly cohesive and resilient popular resistance that neither the state nor federal authorities could suppress for any length of time.

This study raises the level of examination toward the abstract, examining Reconstruction in the South as a struggle between two actors—a dominant actor and a weaker actor. This struggle is asymmetric warfare—a transition from *war* to *warfare* through the mobilization of a collective identity that achieves a mutually acceptable political and social equilibrium from disequilibrium. The weaker actor in asymmetric warfare seeks to attain its goals to prevent the dominant actor from being able to muster its power effectively and forcing the dominant actor to seek a compromise that allows a political equilibrium to arise that, in turn, leads to reconciliation and peace. This approach avoids caricatures and categorization in order to illuminate a process of tactical and strategic resistance that the weaker actor employed to gain its goals at the expense of the dominant actor.

This study attempts to reveal the environment in which the interactions between the actors took place to offer a new perspective that serves as an attempt, at least, at a synthesis of the main currents of Reconstruction writing while also taking a broad-based approach to the scholarly literature on national identity, collective identity, political and other forms of purposive violence, civil wars, reconciliation, and peace building. The result is a study that outlines an explanation for an outcome. It has no moral, but instead intends to explain much about the outcome itself: the rise of the New South which represented a reconciliation that took more than 10 years to bring about and was not fully accomplished until nearly 100 years after the war. Thus, Reconstruction was neither a lamentable and doomed enterprise engineered by a radical tyrannous majority and imposed on a passive population, nor was it the bright and shining, though temporary, triumph of enlightened progress engineered by a coalition of noble freedmen, black activists, and white idealists. There has to be an injection of objective realism and analysis into this period of American history that reveals it as warfare, an asymmetric contest of wills employing both violence and nonviolence that resulted in an agreement on a suitable equilibrium to establish an American reunification.

The experiences of the war in Iraq points the way to a reexamination of what was happening in the South between 1866 and 1877. These 11 years, roughly parallel to the 10-year U.S. experience in Iraq, illustrate the challenges of reconstruction when faced with a mobilized and organized resistance. What our recent experience highlights for consideration of the past is that more often than not, war represents a continuum—peace is not always guaranteed by outright military victory. Depending on the terms the victor imposes, there will follow some level of asymmetric warfare until a state of mutually acceptable political equilibrium is achieved, usually as a result of the exhaustion of the will and resources of one of the actors. In both the American South and Iraq, the immediate political objective of the U.S. government was achieved by military force. The defeated populace accepted the conditions of military victory and the essential changes that came about as a result. For the South, it was the end of slavery; for Iraq, it was the elimination of the Baath Party dictatorship. In both cases, those who accepted defeat expected a return to the prewar norms with minimum interference. In both cases, however, new conditions were imposed by the dominant actor that changed the political and social structure. A new set of unanticipated requirements were imposed and compliance was expected, backed by the implicit use of force. The scope, rate, and direction of imposed political and social change in both cases had a direct effect on the initiation of and expansion of asymmetric warfare. For both Iraqi Sunnis and white Southerners, there came a decision point: to many, compliance was a more costly choice than resistance.

The political leaders of the North failed to understand the nature of the challenge they were facing, and the resulting lack of means to support the strategy allowed white Southerners to mobilize a Southern-Confederate collective identity and gain the initiative though a combination of violent and nonviolent actions that gave Southerners the asymmetric advantage over the more powerful North. From this perspective, Reconstruction in the South represented a sociopolitical reordering at the expense of the weaker actor, former Confederates, who, though vanquished on a conventional battlefield in a state-versus-state clash of wills, were unwilling to accept the terms of the new normal being imposed by the victorious North as the dominant actor. During the Reconstruction period, the white Southerners in each state turned battlefield defeat into a favorable new normal through a transition to asymmetric warfare. A momentary, yet false, equilibrium existed under Presidential Reconstruction, but when Congress implemented a new strategy with ambitious new transformative goals, a condition of disequilibrium emerged. The outcome of Reconstruction in favor of the weaker actor was determined by the employment of asymmetric warfare means that, in turn, had a direct effect on the will and ability of the national and state governments to accomplish Congress's political, social, and economic strategic goals. This decline of will manifested itself in waning popular support for the implementation of the Congressional Reconstruction strategy.

The process of asymmetric warfare, revealed in a series of phases that weakened the legitimacy of the state Republican governments while at the same time gaining the dominance of the information battlefield to weaken the national will, led to a Democratic Party resurgence in the South, the abandonment of its strategic goals in Congress, and a search for some level of compromise and conciliation. These factors combined to provide a new political and social equilibrium that the South presented to the Republican national leadership. In doing so, the Southern states established an acceptable level of a balance of power that created an imperfect, yet stable and politically cohesive region that was now able to advance with the North's tacit acceptance as the New South. Reconstruction was a period of asymmetric warfare—an intermediate state between war and what came to be considered peace. In 1877, it became peace at the expense of a group of Americans who were recognized as free quasi-citizens, but nothing more. But it was still a peace nonetheless—a peace that defined legal, social, economic, and political norms through a mutually beneficial negotiated settlement until the 1970s.

Because so many fine historians have chronicled the course of Reconstruction over the years, I have relied on their historical narratives to examine key events as transition points in shaping the course of asymmetric warfare in the South. It is important to note that the South was not a single political entity, but the establishment of a mobilized collective identity allowed white Southerners to wage asymmetric warfare essentially as a nonstate entity. Examining Reconstruction as asymmetric warfare, applying the assumption that war and identity are completely enfolded, allows a new approach to the tangled web.[9]

Because asymmetric warfare is so heavily dependent on mobilization, I used contemporary newspapers, primarily from the South, but also from the Midwest and border states, and, occasionally, from the major New York City newspapers and even the far West. The newspapers of the South provided readers with a perspective on events both locally and regionally and served as a common source for individuals to forge a common outlook and understanding of their situation that helped strengthen a collective identity that led to mobilization. The newspapers were as much of an asymmetric warfare battlefield as any other in the South, and words were a powerful weapon in shaping opinions and perceptions of what was happening. News items from very wide sources were copied and republished in Southern newspapers, providing a remarkably consistent understanding of activities going on in the South regardless of where the reader was located. The veracity of some of the reporting is irrelevant; what was being read was being absorbed, and what was written revealed the evolution of the four phases of asymmetric warfare. Asymmetric warfare manifested itself in enough disparate locations, at a number of different levels, and at generally the same time that, for many observers, contemporary and modern, it appears as a singular, organized effort. As events take place, actions and reactions occur that create a certain temporary level of initiative for one actor or another; if asymmetric warfare is successful, a transition takes place at some point where the initiative stays with the weaker actor. Newspapers reveal all of these transitions as well as reflections of mobilization and will that provide a sense of immediacy to the narrative.

1

Asymmetry and Asymmetric Warfare

Asymmetry in its most basic form relates to variances, dissimilarities, or inequalities between adversaries. Asymmetry is often understood as a condition in which one actor is significantly stronger than another. Put another way, asymmetry is "the absence of a common basis of comparison" between a dominant actor's and a weaker actor's capabilities, interests, and commitment.[1]

These dichotomies between actors that resist comparison can lead to the weaker actor initiating asymmetric warfare against the dominant actor, an operational method that has been employed by weaker forces against superior forces throughout history. Asymmetric warfare is employed by a weaker actor to engage in a level of resistance intended to prevent a dominant actor from gaining its goals by increasing the level of cost for the dominant actor to an unacceptable level, leading to equilibrium, where the dominant actor accepts a far more limited outcome that conforms to the weaker actor's interests. In the conduct of asymmetric warfare, the weaker actor seeks to gain freedom of action by acting, organizing, and operating differently; and by varying techniques, methods, and approaches to negate the strengths of the more powerful opponent. As the weaker actor seeks to evade or undermine those strengths, it simultaneously seeks to maximize its own relative strengths to take advantage of the opponent's weaknesses and vulnerabilities.[2]

Asymmetries between actors exist in many forms, all of which are a means to an end and play an important role in shaping the course of asymmetric warfare. These categories of asymmetries include asymmetries of power, asymmetries of interests at stake, asymmetries of ideals and culture, asymmetries of organization, and asymmetries of will related to commitment, resolve, and timescale. In every case, the weaker actor seeks to exploit the dominant actor's capabilities in these categories. The weaker actor seeks to apply its own asymmetric advantages in sequence or in combination to counter the stronger actor's ability and will to achieve its goals. To survive, the weaker actor must have an understanding of the dominant actor's capabilities, motivations, and means in order to take advantage of the various asymmetries that are being pitted against each other. Identifying the asymmetries that exist between the actors is critically important in determining both offensive and defensive approaches. An important aspect of asymmetric warfare is the degree to which each actor values the issues at stake in the conflict, especially in determining each side's cost tolerance attached to those values.[3]

Surprise is one of the manifestations of successful asymmetric action against a stronger opponent. If the dominant actor overestimates its own strengths and minimizes or ignores the inherent strengths of a weaker actor, the weaker actor can display a dramatic ability to resist. This often takes the form of group mobilizations, which are organized and supported through existing institutions, networks of individuals, or formal and informal social structures. This type of mobilization and collective action can be of a short or a long duration and can take on a number of violent or nonviolent forms, emerging sequentially or simultaneously, with both psychological and physical dimensions.[4]

Asymmetric warfare, as applied in this study, is considered a subset of social conflict involving value commitments (related to prescriptive norms and beliefs) related to a change in relative power between actors or groups based on the mobilization of a collective identity. This study also defines warfare in its broadest sense, as a contest, armed or otherwise, for political purposes. There is an empirical link between warfare and violence. Warfare is the violent means of restructuring political, economic, and social environments, while also shaping individual and collective actions.[5]

In asymmetric warfare, the weaker actor leverages organization, will, and collective identity against a stronger opponent using a combination of violent and nonviolent approaches to exploit favorable circumstances. These circumstances arise when the weaker actor takes actions that the dominant actor least expects, employing surprise, and presenting additional complications in such a way that an opponent's will to continue the conflict is continuously worn down over time. Asymmetric warfare can unfold in varying degrees and times as part popular movement, part paramilitary activity, and part political action, in which the weaker actor seeks to establish its effectiveness against the stronger opponent and gain freedom of action to seize and maintain the initiative.

Asymmetric warfare is characterized by mutually opposed actions and counteractions employed in an attempt to acquire or exercise power. For each requirement that is necessary for the stronger actor to exercise power, the weaker opponent employs an asymmetric approach to negate or frustrate that requirement. For example, the stronger actor must establish legitimacy in its exercise of power; the asymmetric opponent seeks to decay that legitimacy over time through continuous pressure and dynamic actions to disrupt the administrative structure and demonstrate the stronger actor's incompetence or inability to hold or exercise power. The stronger actor seeks to minimize the asymmetric opponent's freedom of action; the asymmetric opponent demonstrates the inability of the stronger actor to impose its will, while achieving popular recognition and influence. The stronger actor attempts to establish a position of strength; the asymmetric opponent avoids these strengths and uses surprise to exploit weaknesses. The stronger actor seeks to solidify popular support; the asymmetric opponent seeks to negate the stronger actor's ability to provide security to the population.[6]

The purpose of asymmetric warfare is to deter, dissuade, discourage, or defeat the efforts of the dominant actor in order to stop the opponent from achieving its goals. The methods employed are often unorthodox, unexpected, and focused on exploiting the stronger opponent's weaknesses to gain an advantage by raising the risk and cost of the conflict and imposing a disproportionate effect on the stronger opponent's morale and will, forcing the opponent to function in an unfavorable environment where the stronger adversary's advantages become liabilities. This environment exists when the stronger adversary loses the initiative and expends more and more resources in self-protection as both the security situation deteriorates and the dominant actor's political will and cohesion drain away. This leads the dominant actor to face a decision either to accept a much more limited goal than originally planned or to abandon the original goals altogether.

Asymmetric warfare's beginnings arise organically within groups of weaker actors when confronted with common threats and fears, based on value systems, mind-sets, and perceptions, as well as commonly held ideas and patterns of behavior that relate to collective identity. The circumstances and conditions created by the dominant actor generate grievances and hostility at every level and shape the nature of the weaker actor's asymmetric approach, intended to create a level of confusion, surprise, and uncertainty for the dominant actor. Initial resistance is often informal, ambiguous, and fluid as groups seek a position to gain an advantage. Participants can include criminals and outlaws, identity-driven or ideologically motivated social groups, overt and covert organizations, and instrumentalist leaders, as well as legitimate businesses, families, and church leaders. The pulsations of violence that result vary in intensity, are multiple and complex, and interflow with state, local, and national political contests, presenting multiple challenges to the dominant actor that defy definition.

Although there is no coherent plan or central direction, the violence will be most often concentrated in areas where political actors are most vulnerable. The violence takes two forms: in its demonstrative form, violence is employed to demonstrate will and capability, to gain support, and to increase the level of mobilization within the collective identity; in its terroristic form, the violence is employed to deter or prevent political actors from supporting the enemy.[7] This is warfare of action, employing tactics that are both violent and nonviolent applied within a strategic logic and employing symbol, propaganda, and ideas, all sustained by a collective identity that involves a ruthless determination to gain a level of control and influence at the expense of the stronger opponent, usually through marginalization and intimidation of supporters and key political actors.

If the weaker actor gains the advantage in this contest of wills, the stronger opponent will have difficulty using its superior capabilities to any strategic advantage, and the will of the government and its security forces to continue the fight will drain away. The result is that the existing political structure in the eyes of the population is delegitimized, opening the way for some form of compromise and equilibrium to exist that will bring about a new political order favorable to the weaker actor.

PHASES OF ASYMMETRIC WARFARE

Asymmetric warfare depends on a long-term view. By its nature, asymmetric warfare cannot achieve outright victory. Unlike traditional war, asymmetric warfare is about means over ends. Ironically, the weaker actor can never attain its ends without the consent of the more powerful adversary. This disparity between means and ends is what allows the weaker actor to employ asymmetric warfare as the means to wear down the stronger opponent until some level of mutual consent can be achieved. The goal of asymmetric warfare is to create conditions for negotiation to establish a favorable mutual outcome based on achieving a political equilibrium.

Asymmetric warfare has four distinct phases shaped by mobilization of groups through a collective identity, competing group priorities, the level of public involvement and interest, and leadership. These phases represent an evolutionary response and adaptation to existing conditions and the actions and counteractions taken to gain and maintain an advantage in order to exploit the circumstances that emerge in the contest between the weaker actor and the dominant actor. Both timing and time itself are critical factors, influenced by the often-shifting conditions that shape the political, social, economic, and psychological environment of the conflict.

Phase I is Sources and Preconditions for Asymmetric Warfare. This phase emerges from an imbalance or ambiguity of power relations and the incompatibility of sociopolitical goals between the dominant actor and the weaker actor. A new power relationship often is the trigger, as it establishes the comparative strength of the actors. When one actor is placed at a distinct disadvantage to the other, and the more powerful actor threatens to cripple or eliminate the other actor's ability to influence events, the relative weaknesses that are established as a result of this new relationship prevent the weaker actor from having a vested interest in maintaining a stable transition or accepting the new status quo. This reaction to the imbalance generally begins with nonviolent conflict, such as political agitation or public protest; if conditions of imbalance persist, an active resistance emerges employing armed violence. The conditions of asymmetric warfare emerge as the weaker actor demonstrates a high level of interest and resolve to initiate and sustain action, while maintaining a low vulnerability to efforts at suppression by the dominant actor.

Phase II is Identity-Based Classification, Distinction, and Mobilization. This phase marks collective identity formation or manifestation, fostered by interests in reasserting or regaining power. Collective identity is mobilized and demonstrated through commonalities of shared motives, needs, values, attitudes, beliefs, and perceptions. The collective identity of the group is often defined by its opposition to the dominant group, which represents a threat to the survival of the weaker group. This identification supports the cohesion and commitment of the group to achieve its goals. Once mobilized, the weaker actor seeks to gain some level of advantage over the dominant group to demonstrate its effectiveness and then begins to take action in both violent and nonviolent modes. In Phase II, distinct identities and interests are manifested in terms of three actor subsets: the participants (those engaging in nonviolent activity, grouped around a collective identity in which friends and enemies are clearly delineated); perpetrators (those engaged in violent activity as part of an organized group or as individuals who usually, but not always, share the same collective identity as the participants); and targets (those identified as the out-group, based on specific attributes such as race, political identity, or regional affiliation).[8]

In Phase II, with group identity fully mobilized, a threshold occurs as more individuals are willing to take action on behalf of the group in the belief that they can make a difference through their participation. In this phase of asymmetric warfare, violence is the operational instrument that entails the threat of, or actual use of, physical force to protest, resist, influence, or change the environment that favors the dominant actor. Because this rising tide of violence is unpredictable and is both social and asocial, it reflects an unorthodox approach essential to asymmetric warfare as it has numerous modes, representing a number of independent and collective actors, whose purposes and intents are layered and distributed and create a sense of disorder.[9] Freelance violence with political overtones usually emerges. Local violence serves higher political ends, whether or not the violence is actually related to those political ends. The growing level of violence, though not centrally controlled, demonstrates the inability of the government to take effective counteraction and thus encourages more members of the mobilized collective identity to cross the threshold from passive to active support. In general, violence tends to decline when the weaker actor has gained a level of initiative that allows political action to become the dominant mode of influencing the environment.

The efforts of the weaker actor are intended to create uncertainty to a point where the security forces and judicial authorities of the dominant actor are neutralized by an environment of instability, confusion, fear, and psychosocial tensions related to election cycles or particular events that are political in nature. The meanings that the population attaches to these events and activities become as potent (or more so) as the actions themselves. In addition, political groups can ally with local actors to guide or direct violence to intimidate, neutralize, or eliminate political opponents or sympathizers. The community level is the decisive battlefield at this stage. Local actions create a blend of stability and instability and are largely disaggregated and decentralized from the higher political level.

Phase III is the Dislocation of the Dominant Actor. This phase is characterized by psychological dominance over the dominant actor and political freedom of action through the demonstration of highly organized paramilitary capabilities in support of political activities. In this phase, local actions become largely centralized and aggregated at the higher political level. Leaders emerge to challenge the dominant actor for power as the dominant actor reconciles to the now unattainable sociopolitical goals it originally pursued and accepts the attainment of lesser goals. Violence can still exist, in the form of bulldozer tactics, but at this stage, the threat of violence directed at targets has a more important effect, neutralizing and demoralizing opposition and demonstrating to all observers that the dominant actor is incapable of resisting. The shifting of the power relationship from asymmetric to equilibrium begins, as modes of resolution emerge along with new norms that recognize the new status of the weaker actor.

Phase IV is Equilibrium and Reconciliation. The conflict ends with the establishment of a political equilibrium—a consensual framework satisfactory to both sides. An acceptable political alternative is established after the formerly dominant group recognizes that it lacks the will and means to achieve its original goals. The factors that initiated asymmetric warfare are resolved, and the new political and social environment is characterized by functional and stable relations within a desired range of change acceptable to both actors that leads to reconciliation and the path to peaceful relations.[10]

The preceding outline indicates that asymmetric warfare is not guerrilla warfare, insurgency, or terrorism, although these terms are often used synonymously with asymmetric warfare. Indeed, all of these terms have been used to describe the violence that occurred in the Reconstruction South, and all of them are inaccurate. Guerrilla warfare and insurgency share the same technique of using armed force in a protracted struggle to overthrow the state and replace it with a completely new political order. In an insurgency and in guerrilla warfare, the focus of effort is armed resistance, where military actions are subordinate to strategic political objectives. Guerrillas operate in small formations organized along military lines and use unconventional tactics to attack the lines of supply and communication of conventional military forces to render those forces unable to prevent the final overthrow of the state. Insurgents target security forces and use subversion and violence to challenge political control as they seek to gain popular support to destabilize or overthrow a government. The focus of both groups is on the military defeat of an enemy conventional force by employing hit-and-run tactics to avoid direct battle whenever possible while seeking to gain support of the population and turn them to their cause.

Mao Zedong described guerrilla warfare as people's war. The guerrilla sustains himself through the people, who provide shelter, food, and other resources. The people have to be mobilized through cadres of dedicated professional revolutionaries to accept and support the guerrilla's cause. In the Maoist concept, guerrilla warfare is only one stage in a progression to conventional mobile warfare to gain eventual victory, thus not an end in itself. Although guerrillas and insurgents may use terrorism as a method, they themselves are not terrorists.

This study's approach to asymmetric warfare differs from the separate categories of guerrilla warfare and insurgency in that the goals and actions of asymmetric warfare are significantly different in purpose, nature, and scope from the goals and actions of guerrillas or insurgents. Asymmetric warfare differs from guerilla warfare and insurgency in that the sources of mobilization are tied to issues of identity. Conventional military forces are not the target in asymmetric warfare, and, unlike guerrilla warfare and insurgency, the overthrow of the state through military victory is not the objective. Asymmetric warfare is multi-leveled and functions effectively at the local level, where secret societies, bandits, and paramilitary organizations can proliferate. In asymmetric warfare, there is little need to propagandize and win the hearts and minds of the population because a significant part of the population already is either actively or passively engaged. All are mobilized and sustained to one degree or another through a collective identity that provides the asymmetries of will and time that are essential to a successful outcome.

Terrorism, employed in the nineteenth century sense of targeting symbols of political authority, was used both as a means of propaganda and as a symbolic attack on the state. In this sense, terrorists do not seek the support of the people or to change the government; instead, terrorists seek the people's acceptance of their agenda.[11] Unlike terrorism, the violence in asymmetric warfare is decentralized; often focused at the individual, group, or even regional level; and is usually associated intentionally or otherwise to a wider political confrontation that serves the interests of both local actors and central political actors. The intended outcomes of this violence are intangible and often difficult to identify, but the violence does serve its intended effect of demonstrating the power of the weaker actor to counter the dominant actor's strengths and impose increasingly higher costs to wear down the dominant actor's resolve and will. In turn, the effective employment of violence advances the weaker actor's sense of purpose and belonging, and it strengthens the collective identity to continue the resistance.[12]

In the prosecution of asymmetric warfare, the weaker actor exposes civilians to violence because of their importance as political actors. In asymmetric warfare, the home front becomes the battlefront as the entire civilian population becomes enmeshed in the conflict. The weaker actor must sort out participants, supporters, and neutrals in the search for appropriate targets.[13] Violence is a resource for the weaker actor because of the simultaneous strategic and tactical effects it has on the target group. Civilian deaths or injuries serve the goals of the weaker actor by pressuring the target population into passivity and political abdication and shaping popular perceptions that contribute to weakening the government. Violence is also a resource for mobilization of the collective identity by building support and incentives for collective action.[14]

Violence against civilians takes the form of noninstrumental or indiscriminate attacks, usually to displace people or to demonstrate power and capability; it is more commonly employed in instrumental or selective ways as a means to shape or coerce a desired behavior in the target group and also as a means of control, giving targeted individuals the option of shifting support to the weaker actor. Whether noninstrumental or instrumental, the violence serves a communicative function—collectively the target group is equally vulnerable as any individual member of the group. Physical violence takes a number of forms, including arson, vandalism, forcible displacement, beatings, torture, mutilation, rape, desecration of dead bodies, and murder. As the essential component for the prosecution of asymmetric warfare, violence assumes a layered character, exhibiting political, local, regional, and personal elements intended simultaneously to intimidate and demoralize, demonstrate and sustain morale, and support mobilization. It also serves to establish and maintain social control and community standards.[15]

THE CENTRALITY OF COLLECTIVE IDENTITY
IN ASYMMETRIC WARFARE

Identities develop and evolve through interactions within a society and are tied to self-definition. Collective identity gives individuals a sense of place, purpose, and direction, as well as a measure of psychological security, which is integral to the construction and maintenance of identity. Collective identities are manifest in many ways, through symbols, language, values, history, and a set of cognitive beliefs, understandings, and norms.[16] Attachments and commitments are made through a narrative, which collects ideas, representations, and emotions into a mental image that serves to explain the group's past and present, as well as the status and condition of the group as a means of collective identification. These narratives are central to collective identity.[17] When identities are linked to a political purpose, they become fixed and formidable. When this happens, individuals will subsume their own identities within the collective identity and see a threat to the group as a threat to their person. Linked by personal and social relationships with other members of the group, strong bonds are formed. When social and political conditions force individuals to choose one identity over another, they will react to protect and secure their threatened identity, often reacting violently in pursuit of collective interests or defense of values.[18] Individuals mobilize identity within specific political, economic, or social contexts by supporting or creating organizations, "channeled through existing state and social structures, processes, and institutions."[19] Once mobilized in protection of that identity, the group will view itself as distinct, separate, and unique, coalescing around what defines the collective identity, while also identifying the characteristics and markers of difference of those who do not share that identity. This is the out-group, considered the threat to the survival of collective identity. When the structures and process are in place for mobilizing identity, the group seeks to gain the psychological initiative, supporting their cause under any circumstances to present an image of strength and credible power. At this point, collective identity is consolidated, the polarization between groups heightens, and conflict becomes inevitable.[20] Thus, in asymmetric warfare, security and identity merge in reaction to real or perceived threats that involves mobilized actors of distinct identities pursuing interests through violent confrontation. This mobilization contributes to asymmetry of will and interest and gives the weaker actor a distinct advantage.[21]

CONSIDERATIONS: THE NORTH CONFRONTS
THE AFTERMATH OF CIVIL WAR

The Civil War was a classical symmetric conflict: the armies of the Union and Confederacy were similarly organized and equipped, using the same tactics, and were employed in campaigns to achieve similar objectives. Thus, a symmetric war "is a conflict between well-matched forces, where similar technologies, capabilities and vulnerabilities exist, and tends to be attritional, costly, and long-term."[22] The Union army dictated the type and course of the war because it had the ability to continue waging it by effectively employing the massive resources at its disposal. This total war approach stretched the Confederacy to its breaking point. The hungry, worn out men who surrendered with Robert E. Lee at Appomattox and Joseph E. Johnston at Bentonville were quite a shocking contrast to the well-fed and well-equipped troops of Ulysses S. Grant and William T. Sherman. There was no doubt in the minds of the commanders that the means of organized resistance had been eliminated. The overwhelming military victory convinced Northerners that the South was indeed decisively beaten and incapable of further resistance, opening new pathways to transform social relations and create new distinctions. For the four terrible years of war and, arguably, for many years earlier, Southerners had been demonized, first as an evil slavocracy that was a cancer in the American body politic and then as traitors and rebels who had attempted to destroy the Union for selfish gain. The emergence of a racial minority as a new element of freedom in the life of the nation had played a central role in the war and was considered one of the fruits of victory. Military force and public resolve had achieved the Union's war aims. The victory won at such a cost in human lives justified the outcome and allowed the North to create its own narrative of virtuous triumph over evil, sanctifying the victor while demonizing the vanquished.

Now possessing unchallenged legitimacy, the monopoly of force, and the opportunity to exercise its power through democratic means, the North faced an immense task, represented by the multiple challenges of economic development, institution building, and social reform, to create a new South through political, ideological, and cultural reconstitution and readjustment. A number of unprecedented challenges faced the victors. The social fabric of the slave society of the South had been destroyed. The slaves were indeed free as a result of military action, but the actual meaning of freedom for these people was still not fully determined or understood. The Union armies were confronted with massive social dislocation combined with an unprecedented level of human suffering. By military necessity, the economic infrastructure of the South had been utterly destroyed. Beyond dealing with the immediate necessities of occupation, the military and political leadership of the North now had to determine what combination of social, political, and economic reconstitution and readjustment would bring about a lasting peace, while at the same time achieving the lofty goals of the Republican Party that now had the power to reshape the nation forever in its own image.

Yet the social and political factors that gave rise to the war in the first place had not been resolved by Union military victory, and the problems arising from those factors now confronted the victors in the aftermath of war. Although the Southern people's motivation to resist during the war had never been addressed or appreciated, there was some consideration for establishing a level of control of the political environment of the defeated South that was necessary to avoid a reemergence of an organized threat in the future. This in itself was a complicated problem: while state governments still existed, no central government authority remained. It was clear that government institutions and processes would have to be restructured, but to what degree and how were as yet not fully appreciated. The humanitarian concern for the freed people and white refugees required that they had to be provided with a level of security and support to begin to reestablish some immediate level of social order and stability. With the cotton economy shattered, some process had to take place to build up to some semblance of productivity through capital investment and redevelopment that would allow a new economy to emerge that conformed to the North's economic interests. In essence, some level of political equilibrium was necessary if any progress was to be made.[23]

A number of nagging and uncomfortable questions arose beyond the immediate security, humanitarian, and economic concerns. What was the future of former Confederates in the new South? Who would be included, and who would be excluded? Could the effects of the war that had separated the bonds of union between the sections be mitigated to the extent that they would not interfere with future development? With the war won, a different level of commitment and cost tolerance would be needed for the North to achieve its goals. The essential question was this: could the same organizing power of the state, exercised in war for the salvation of the Union, be put to use in establishing peace?[24] How the political leaders in Washington, D.C., recognized and understood the totality of what confronted them would determine the nature and direction of postwar America. Because of their overconfidence in their own virtues and because of the overwhelming power they possessed, these politicians failed to grasp the full consequences of their actions and opened the way for a transition not from war to peace but from war to asymmetric warfare.

Without equilibrium, conventional military victory only marked the end of one form of the larger conflict. As a result of the unacceptable changes in the postwar environment, the conflict itself changed shape and form as white Southerners made a transition from conventional war to asymmetric warfare. As warfare, violence is implicit and necessary to advance the weaker actor's interests and goals.[25] Violence in warfare is used to attack will and capacity to fight by wearing down the opposition to neutralize opposition armed groups and avoid regular forces. Through this violence, the weaker actor must demonstrate an ability to undermine the state governments by creating threats that the state is incapable of addressing. When violence is employed in this manner, it will continue until the actors agree to some type of settlement and reach a political equilibrium.

Thus, the story of Reconstruction in the South is about this transition to asymmetric warfare and its employment in four phases to arrive at a final political equilibrium that leads to the process of reconciliation. Understanding the context and contours of asymmetric warfare places the Reconstruction period in an entirely new light. Clearly, in the complex social and political postwar environment, the South was a weak actor and had ostensibly no means to resist whatever new conditions should be imposed by the dominant North. But the South had several asymmetric advantages that the victorious North overlooked. The underlying social conditions of group heterogeneity, combined with geographic concentration, and collective leadership that had made it a formidable enemy in war, were still quite strong in the defeated South.

These conditions were enhanced by a Southern collective identity that was essential for asymmetric warfare. This identity manifested itself in group resolve, a social structure that had been hardened by war and privation and inured to violence, and allowed white Southerners to mobilize, organize, and take action in the face of what they perceived as a total and final annihilation of their place in the new America. This identity, based on a shared culture and the Confederate national self-perception of honor, resilience, and sacrifice, contributed to the eventual cooperation among disparate groups within the South to join in opposition to the Radical plan of Reconstruction. The inherent nature of asymmetric warfare in the Reconstruction South was political, not military. It was a melding of identity and action. It represented a transition from civil war to civil struggle, that is, a war of pure military force to warfare involving words and ideas and punctuated by the violent hostility of a mobilized collective identity.[26] The struggle itself was highly symbolic and took place on several levels—political, social, and private. The North never recovered from its shock and surprise at the South's reaction to the new conditions imposed. As its vulnerabilities continued to be exploited, the dominant actor became less effective in using its superior power. More resolute and less politically vulnerable than the North, the American South's conduct of asymmetric warfare was unusual. It was not a united effort, but a patchwork of separate state and local level political-military warfare activities, each unique, but all sharing a common will and common goals. By combining aspects of peace and violence in unexpected and unanticipated ways over nearly a decade to impose higher and higher political and durational costs on the North, a political equilibrium that conformed to the interests of the white South was inevitable.[27]

2

From False Equilibrium to Disequilibrium: The Strategies of Reconstruction (1865–1867)

President Abraham Lincoln did not outline a strategy for Reconstruction. He had, instead, an intent that was based on the two political and strategic goals of the war: a restoration of the Union and the abolition of slavery. Military occupation of Tennessee, Louisiana, Arkansas, and parts of South Carolina and Virginia gave him wide latitude as a wartime commander in chief to pursue his intent. Because Lincoln believed that the states had never left the Union, the only necessity was to restore a loyal government. Although he did not provide any specific guidance either to the military governors of the occupied areas or to the military commanders conducting occupation duties, Lincoln signaled that local conditions would dictate how the political and military authorities would cooperate to restore legitimate government authority under his general plan. For Lincoln, restoration of the Union required Southern loyalists, mainly former Whigs, to form the nucleus of a state government that would acknowledge the new reality imposed on the former Confederate states by force of arms: that the institution of slavery was dead. After loyal control was restored, Lincoln believed the issue of the rights of the freedmen would be addressed by each state according to its own situation.[1]

Lincoln's 1863 Proclamation of Amnesty and Reconstruction outlined a restoration of state governments that required 10 percent of voters to take a loyalty oath. Included was a blanket amnesty for those taking an oath of allegiance and full pardons from the president were offered to military and civilian officers of the Confederate government who applied. With the war's end certainly in sight, the president approved the joint resolution of Congress submitting the Thirteenth Amendment to the states for ratification in February of 1865. With these two political strategic goals met, Lincoln's strategic intent would also be accomplished. The partnership between loyal Southerners and Union occupation forces would be capable of addressing the details of restoring a functioning government without direction or oversight from Washington.[2]

This general outline was behind Lincoln's meeting with his key military commanders at City Point, Virginia, on March 27, 1865, to discuss how the war was to end. According to Gen. William T. Sherman's recollection, Lincoln wanted the war to end as soon as possible to avoid further bloodshed. Although he did not communicate his strategic intent, Sherman was certain that Lincoln had in his mind "some plan of settlement ready for application, the moment Lee and Johnston surrendered."[3] Admiral David Dixon Porter also recalled that "Mr. Lincoln came down to City Point with the most liberal views toward the rebels. He felt confident that we would be successful and was willing that the enemy should capitulate on the most favorable terms."[4]

In the aftermath of the fall of Richmond on April 3rd and the evaporation of the Confederate government, Assistant Secretary of War John A. Campbell questioned Lincoln on his desired conditions to end the war. Lincoln provided a memorandum with three goals: the disbanding of Confederate forces currently in the field, the restoration of the authority of the national government, and the end of slavery. On April 7, 1865, Campbell summarized the president's intent in a message to Maj. Gen. Godfrey Weitzel, who was functioning as the commander of occupied Richmond. The three conditions, he informed Weitzel, "which when examined are [all] included in the single one of the restoration of the Union by [the] consent of the seceded states."[5] Weitzel's request for guidance may have originated with the general's encounter with the president on April 4th, during his visit to Richmond. He asked President Lincoln what his orders were for dealing with the people of the city he now controlled. Lincoln refrained from giving any orders but signaled his intent, telling the general, "If I were in your place, I'd let 'em up easy, let 'em up easy."[6]

As the war was ending, Lincoln had only conveyed a very broad strategic intent, seeking to end the agonizing bloodshed and destruction of the past four years and restore peace as quickly as possible. The next step in accomplishing his intent was establishing loyal state governments that accepted national authority, reflecting Campbell's paraphrase of restoration through consent of the states themselves. The slavery question, temporarily resolved as a wartime measure, was adequate for the time being. White control would remain, and the future condition of the freedmen would be addressed later. Lincoln's death, a few days after his conversation with Weitzel, left a critical gap in leadership to guide further strategic intent for the postwar South.

In contrast to Lincoln's lack of strategy, the leadership of the U.S. Congress had been in the process of formulating a postwar strategy for several years. To the Republicans in Congress, the purpose of the war was to suppress a rebellion of traitors who created and defended a Confederate States of America in defiance of the Constitution. The pursuit of total war to destroy this rebellion included emancipation of slaves held in the rebellious states as a weapon of war to deny the Confederacy the means to continue resistance. As the Union armies advanced and occupied territory, the military necessity of dealing with the large number of former slaves who were now free under military authority became a growing concern. The War Department created the American Freedmen's Inquiry Commission to address the consequences of emancipation and assess the means needed to provide for the protection and improvement of the black population within friendly lines. In May of 1864, the commission rendered its report, recognizing "the base prejudice of our race to a colored man," but nonetheless outlined requirements for basic humanitarian needs and also a recommendation for providing advice, support, and educational opportunities to inculcate self-reliance and productive employment.[7]

In the same month, another report came from the Congressional Committee on Emancipation. The committee was chaired by Senator Charles Sumner of Massachusetts, one of the most virulent opponents of slavery and a powerful voice in Congress. The report established the outlines of a postwar strategy. First and foremost, slavery would be ended, indicated unequivocally by the Senate's recent passage of its version of the Thirteenth Amendment. The rebels were to have their property seized, and all claims and debts would be invalidated. The Confederate states would be treated as a conquered territory. The land of former slaveholders would be redistributed to free blacks and poor whites. The black population, considered to be powerless without some means of protection, would be afforded that protection by granting them civil rights, including the right to vote. By making them citizens and affording them equal rights, blacks would assist in the restoration of the Union. This transition would be supported by an organization, backed by the federal courts, that would protect the new rights of the freedmen, ensure equal justice, and offer economic and educational opportunities to create a population of self-reliant workers and farmers. The recommendations of the American Freedmen's Inquiry Commission established the Bureau of Refugees, Freedmen, and Abandoned Lands (the Freedmen's

Bureau) in March 1865, signaling Congress's interest in addressing the civil rights of the freedmen and continuing postwar relief efforts through the Freedmen's Bureau. At the same time as President Lincoln was articulating a broad and general intent, the Republican Congress was already uniting in support of a clear strategic outline of military occupation, punishment of traitors, vast economic reform through land redistribution, and an extraordinarily ambitious program to constitute an entirely new group of citizens fostered under the guidance and protection of federal power.[8] Implementing this strategic outline, however, would have to wait until the new Congress convened in December.

Lincoln's violent and tragic death darkened the national mood significantly, rekindling the collective memory of 300,000 Union dead at the hands of rebels and the public anger over the loss of the thousands of Union prisoners who died under inhumane conditions.[9] Those who had supported Lincoln's conciliatory approach to the South now believed that Confederate leaders had conspired to murder the president and spoke of punishing treason and meting out just retribution.

Hostile crowds seeking some form of revenge formed near Fort Warren where Gen. Richard S. Ewell and other Confederate generals were confined outside Boston. Even as far away as California, furious crowds vented their anger. Henry George, the future author of *Progress and Poverty*, was working as a substitute typesetter for a newspaper in San Francisco. When the news of the president's assassination became known, crowds attacked pro-secession newspaper offices, wrecking them wholesale. George had formed a gang of his own, intending to settle accounts with the disloyalists, but was too late to do any additional damage. The following day, he wrote a eulogy to Lincoln that was printed in his newspaper and captured the level of hostility directed toward Southerners at this time:

> The murderer's shout as Lincoln fell [John Wilkes Booth had shouted "*Sic semper tyrannis!*" after he jumped from the president's box at Ford's Theater], it will be taken up by a million voices. Thus shall perish all who wickedly raise their hands to shed the blood of the defenders of the oppressed, and who strive, by wickedness and cruelty, to preserve and perpetuate wrong. Their names shall become a hissing and a reproach among men as long as the past shall be remembered.[10]

From that sad April until December, a new president had a significant challenge. Andrew Johnson was a staunch Tennessee Unionist and Democrat, a self-made man, and a rough political brawler. But he did not dwell in "the province of enlarged and wise statesmanship" required in the uncertainty of postwar America.[11] Professing a deep hatred for the slaveholding elite, which he held responsible for bringing the nation to disunion and war, he also was equally as strongly attached to the common man and shared his aversion to the elevation of the freedmen. He viewed emancipation as a wartime expedient and shared Lincoln's interest in restoring the states using a core of loyal voters. However, Johnson's view was that these voters would serve as the base for a revitalized Jacksonian democracy, not the Republican concept of freedom, equality, equal rights, and free labor.[12]

Johnson was initially welcomed to Washington as a man who would chastise the South and satisfy the North's passion for revenge. Despite his often bloodthirsty rhetoric during the war as the military governor of Tennessee and his friendly attitude toward the Radicals in Congress, Johnson sought to follow Lincoln's strategic intent for a rapid restoration of the Union. Wasting no time, he began the process in May of 1865. The state governments of Virginia, Arkansas, Louisiana, and Tennessee, which had been structured according to Lincoln's intent, were declared legitimate. For all other states, Johnson appointed a provisional governor and required that a state convention of voters who had taken the amnesty oath would then establish the procedures for establishing a government. Pardons were available to the planter aristocracy—an added restriction imposed by Johnson—and the military and political leaders of the Confederacy. This new government would declare the ordinance of secession as null and void, repudiate the Confederate debt, and finally ratify the Thirteenth Amendment abolishing slavery.

Johnson's approach was to get Reconstruction out of the way, establish a new equilibrium, and proceed forward as a reunited nation. His requirement for the ratification of the amendment abolishing slavery was his attempt not only to recognize the achievement of an essential political strategic goal of the war but also to demonstrate to Congress that the newly constituted postwar state governments had accepted this condition as a requirement for reunion. Under Johnson's plan, the Southern states were allowed to order their own affairs as they saw fit without further federal oversight. The states proceeded to deal with the freedpeople just as they were now under the newly amended Constitution—free people who were neither citizens nor property, and, as such, the states placed them under restrictions intended to maintain order and control while acknowledging some legal rights and restricting others.[13] In the restoration of state governments, Johnson sought a Jacksonian revival of true democracy. Instead, the voters returned political power to the prewar ruling elite who Johnson believed had led them to secession and war. The Southern states believed that within the bounds of the presidential intent for reconstruction, restoration of federal control alone was sufficient for reunion. A South Carolina newspaper outlined the argument. Congress had declared that the war was waged "not to deprive the 'rebellious States' of their rights, but purely and simply to restore the authority of the Constitution and laws of the United States." With the war over, that result had been achieved, and the federal laws and institutions were functioning under the authority of the Constitution.[14]

The newspaper's pronouncement captured the president's thinking. Johnson's achievement in reestablishing the authority of the Constitution and the laws of the United States represented a political equilibrium that existed only between the president and the former Confederate states. Johnson had gone forward to fulfill Lincoln's intent, and he considered the state elections as the main contributors to stability and establishing a new order; however, he did so without any sense of a national strategic direction or purpose—certainly nothing that he could articulate. Johnson was entirely dependent on the states to take the action necessary to order their affairs in a way that would be acceptable to the rest of the country. Instead of promoting stability, the results of the elections in the Southern states actually contributed to further tension and conflict.[15] Lacking any political adeptness, Johnson carried on without seeking the assistance or advice of the Congressional leadership, many of whom were a proud and severe lot, unaccustomed to being ignored. Having no strategy of his own to articulate to Americans and failing to appreciate the mood of the North, President Johnson's fatal mistake was his faith in relatively simplistic expectations to solve a complex problem in just a few months. As a result, Johnson's newly established political equilibrium was about to be shattered.

The Congressional strategic outline had a clear strategic goal in mind for postwar America—to remake the South in the North's image. Congressional leaders had already defined the ways to achieve this goal in terms of black political (and social) equality, and they had the means through legislation to accomplish it. The new political equilibrium of President Johnson was unacceptable, and when Congress convened on December 5, 1865 (the day before the Thirteenth Amendment became part of the Constitution when Georgia ratified it on December 6), the Southern states were refused representation. By denying the Southern states political power, Congress could now implement its strategy without any interference or resistance from Southern Congressmen. Congress required several precursors in order to move forward. The Thirteenth Amendment, ironically, was an obstacle. Blacks, now free, were no longer three-fifths of a person for apportionment. This meant that simply by population alone, the Southern states would have a significantly larger proportion of representation than they did before the war. Moreover, white Southerners were not natural Republicans and could be counted on to elect representatives and senators who would oppose the Republican strategic goals. Northern Democrats also represented a small, but significant, bloc that would naturally ally with Southerners and reduce the ability of Republicans to pass legislation and override a presidential veto if necessary. Therefore, the only way for the balance of political power to shift reliably to the Republicans to ensure strategic success was to create a loyal Republican voting majority in the Southern states. This required a simultaneous process of reducing the influence of white Southerners who had supported the Confederacy, while elevating Southern blacks through citizenship and enfranchisement to serve as the genesis of the new South. This approach served two purposes close to the hearts of many Republican leaders in Congress. First, giving blacks the vote, while denying the same to rebels and slaveholders, was intended as a sign of righteous retribution; second, it had the highly satisfactory outcome of achieving the Republican strategic goal of a biracial, socially mobile society of property owners, composed of citizens possessing equal rights and fulfilling the free labor ideology of individual self-sufficiency and progress through a commitment to hard work. This inherent dignity of work embodied in free labor would establish the moral and social conditions that would end Southern distinctiveness over time and allow a suitable Northern identity to emerge, creating the optimal conditions for Northern entrepreneurs to build a new South in the image of the North.[16]

Congress established a Joint Committee on Reconstruction in December of 1865 to gain a clearer picture of conditions in the South beyond what President Johnson was portraying with his reconstructed states. By June of 1866, the committee had prepared its report.[17] The committee members, after interviewing over 140 witnesses and asking questions closely related to the Congressional strategic goals, established several broad operating concepts from which to pursue its strategy for Reconstruction. The first and most salient point was that the people of the South, by their act of secession and formation of the Confederate states, were to be considered "insurgents, rebels, traitors, and . . . enemies of the United States." They yielded only when they no longer had the capability to resist. Now as defeated enemies in war they were therefore completely subject to the will of the conqueror. Moreover, as a conquered people, they had not shown any "evidence whatever of repentance for their crime, and expressing no regret, except that they had no longer the power to continue the desperate struggle." As a result, the federal government had every right to seek compensation for the damages incurred to the nation and be assured that such a rebellion could never happen again. Thaddeus Stevens, the Radical Republican congressman from Pennsylvanian, reflected the tone of the committee's conclusion in September when he referred to the defeated Southerners as "the wicked enemy" who had waged an "unjust war" against the Union. That act of rebellion, he believed, must be paid for by changing "the whole fabric of Southern society."[18]

Therefore, readmission of the states formerly in rebellion should only be granted only when "certain conditions and guarantees" were fulfilled. It was up to Congress to determine how the former Confederates would present satisfactory proof of allegiance, as well as determining how long it would take before Congress would restore full "political rights and privileges" to the conquered people. No details were specified, only that these decisions would "depend on grave considerations of the public safety and the general welfare."[19]

Second, Congress asserted that it alone had the constitutional authority to restore the states to the Union and accept representation in Congress. As commander in chief, the president could only constitute temporary governments under his wartime authority. Furthermore, the president had no constitutional authority to recognize state governments as legal with individuals qualified to appoint senators and representatives. None of the state governments newly reconstituted under presidential authority had submitted their constitutions to the approval of the voters. Moreover, the elections held under the president's plan "resulted, almost universally, in the defeat of candidates who had been true to the Union, and in the election of notorious and unpardoned rebels, men who could not take the prescribed oath of office, and who made no secret of their hostility to the government and the people of the United States." The committee regarded the Southern state's acceptance of the president's conditions for readmission to the Union as a mask for gaining admission to Congress so that the war could be continued by political means.[20]

Third, the people of the South displayed no evidence of loyalty, no contrition for their actions, demonstrating both "an intense hostility to the federal Union, and an equally intense love of the late confederacy." There existed a generally intolerant attitude toward anyone friendly to the Union. "Officers of the Union army on duty, and northern men who go south to engage in business, are generally detested and proscribed. Southern men who adhered to the Union are bitterly bated and relentlessly persecuted." There existed a feeling of "vindictive and malicious hatred" toward emancipated slaves. The presence of federal troops was often the only means of protection for many Unionists.[21]

Finally, with the abolition of slavery by constitutional amendment, it was necessary to secure the civil and political rights of the former slaves as "free men and citizens." The committee concluded that "appropriate constitutional provisions" were necessary so that "political power should be possessed in all the States exactly in proportion as the right of suffrage should be granted, without distinction of color or race." The constitutional changes would, in the committee's opinion, establish the conditions of permanent peace, security, and safety that would allow the states of the former Confederacy to be restored to the Union. The amendments would "determine the civil rights and privileges of all citizens in all parts of the republic . . . place representation on an equitable basis . . . fix a stigma upon treason, and protect the loyal people."[22]

The committee's report provided the conceptual outline for Reconstruction, and while it made some very accurate assessments, it often drew the wrong conclusions from the evidence. If President Johnson made the error of taking a simplistic view in attempting to restore the states to their previous status quickly in accordance with Lincoln's strategic intent, Congress was equally in error in its simplistic approach to pursing the strategy. The residual wartime enmity demonstrated in the declaration of Southerners as traitors who needed to pay for their actions was combined with the victor's sense of triumphant power is clear. In Congress's view, total war had reduced the South to ruin and what remained was a conquered people subject to the will of the victor. This assumption, that Southerners had no choice but to accept the conditions dictated to them, was central to pursuing the strategy. It was also absolutely false. Even though the committee identified an intense love for the Confederacy among the people that was at least as equally as strong as the antipathy toward the conquerors, there was no appreciation for what that meant as a threat to achieving the goals of the strategy.[23] The second assumption essential to achievement of the strategy was that constitutional amendments would serve as an all-in-one solution, covering civil rights, citizenship, humiliation of the rebels, and protection of loyalists. The ratification of these amendments was assumed to be the most direct and efficient route to bring about the required level of contrition, compliance, and renewal necessary to readmission to the Union. Another assumption was that Congress controlled the time it would take to restore the Union. The committee only vaguely outlined the conditions in the Southern states that would meet Congressional goals, implying that they would be determined at some future time. The committee noted that because Reconstruction was "a matter involving the welfare of the republic in all future time," it could not be executed "without fully understanding all its bearings and comprehending its full effect."[24] The committee's report unfortunately displayed a lack of understanding of the full effect of the strategic approach of the Congressional Republicans. To a large extent blinded by hostility to the president, outrage at the defiance of traitors and rebels, concerned about racial violence, and consumed by achieving their high ideals that had been validated by so much blood and misery, the committee's report formulated a number of deeply flawed assumptions that formed the basis of the strategic plan for Reconstruction.

Clear indications existed for reconsideration and reassessment, but they were never addressed. At this critical moment, there existed an opportunity in the aftermath of the Civil War that may have changed the direction of history. Congress, under the influence of more moderate and liberal elements of the Republican Party, could have taken advantage of the strategic opportunity presented by the democratizing environment established under Presidential Reconstruction and the tenuous political equilibrium that existed as a result. As the dominant actor, Congress could have used its constitutional power to open a dialogue with the newly reconstructed Southern states to reach a negotiated compromise regarding what constituted a just legal and social order in postwar America. Authority for Congressional representation could have been offered as part of the negotiation and presumably, and the status of the freedmen in America could have been debated, leading to some type of compromise. With such a compromise, a new equilibrium could be established and a path to reconciliation opened that would have led to a stable peace.[25]

But such a result could only come about if the dominant actor recognized that a practical solution is preferable to continued violence and unrest. Tragically, there was no such recognition. The Republicans in Congress, confident of their perception of overwhelming power, decided upon a direct military approach to achieve a rapid transfer of power to Republican-led governments, which would be accomplished with the acquiescence of now largely powerless white Southerners to the new political and social status quo. The committee's report reflected the potentially fatal flaws of a domi- nant actor. Filled with a sense of the triumph of righteousness, possessing unchallengeable military power with its occupation forces in the South, and buoyed with the optimism that came from having taken the initiative to control the future of the country from the executive branch, Congress held the initiative. It had high ideals as its strategic goals, it had a contemptible supine people not thought of as legitimate Americans subject to its will, it commanded enormous power, and it had all the time in the world.

3

Asymmetric Warfare Phase I: Sources and Preconditions (1866–1867)

The Republican majority viewed political change as essential to achieving Congress's strategic goals. They recognized the need to reestablish the legal authority of the U.S. government over the people who had declared allegiance to the Confederate States of America and to delineate their relationship to the national government. Thus, the Fourteenth Amendment was a strategic as well as a political instrument intended to restructure the existing postwar legal-political order in order to carry out the Congressional strategy.

Liberal Republicans joined with Radical Republicans in advocating for universal suffrage for the freedmen. It was fully in line with the Republican vision for America—the suffrage would transform the freedmen into independent citizens, who would be capable of taking care of themselves and who would not need the federal government's assistance or protection. Enfranchisement was, therefore, the most rapid and efficient means to achieve the goals of the Congressional strategy. Once enacted, it was assumed that the amendment would complete the work of Reconstruction.[1]

Serving to provide federal protection for life, liberty, and property, and directed at the sovereign authority of the states, three sections of the amendment followed the Reconstruction Committee's recommended strategic outline. In Section 1 of the amendment, individuals declared free under the

Thirteenth Amendment (former slaves, previously defined as property and existing outside the political community of the South) were made national citizens, and states were prevented from attempting to infringe upon the privileges or immunities of citizens. Section 2 indicated that participation in rebellion was cause for restricting voting rights. Section 3 disfranchised Confederates who had held federal office prior to secession, who had engaged in rebellion, or who had given aid and comfort to those in rebellion. To ensure that the states completely repudiated secession, Section 4 of the Amendment invalidated the Confederate debt. Although the amendment did not explicitly extend voting rights to the freedmen, it did place penalties on states that did not allow freedmen to vote and met the intent of enfranchising free blacks while disenfranchising former Confederates and eliminating the leadership from power.[2]

On June 22, 1866, President Johnson submitted a message to Congress announcing that the Fourteenth Amendment had been sent to the states for ratification. Johnson made it clear in his message that he was only fulfilling his duty, and it represented no support for the amendment. He indirectly defended his reconstruction plan, observing that as a result of Congress denying representation to the Southern states, the people of those states had no voice in ratification. The Charlottesville (Virginia) *Chronicle* reflected the widespread attitude of Southerners. "The Southern people want rest. To secure it they would accept the constitutional amendment." In accepting the amendment, however, the South would be assured that Congress "shall never meddle with the elective franchise in the States, and that the Union be completely restored at its adoption . . . and let the past be buried." As much as Johnson and Southerners wanted the past buried, it was the future that mattered far more.[3]

The Fourteenth Amendment was intended to make permanent the Civil Rights Act passed over President Johnson's veto on April 9, 1866. The concepts of citizenship were central to the Civil War and represented one of the strategic goals of the Republican Congress.[4] It had defined citizenship in the same way as the Fourteenth Amendment and gave citizens the full and equal benefits of the laws in every state and territory. More importantly, all federal officers, including officers and agents of the Freedmen's Bureau, were given arrest and imprisonment authority for violations of the act, and U.S. district and circuit courts would try offenders. The president, or another delegated authority, could at his discretion, use state militia or federal military forces to enforce compliance with the law.[5]

Reconstruction under Congressional oversight and control would realign the South's political structure as well as its social order. The South itself would be remolded in the image of the North, bonded both to the Republican Party and free labor capitalism. A spirit of optimism infused the Republicans, who now saw the road open to ensure a complete moral victory over the defeated South. Pennsylvania Representative and leading Radical Republican Thaddeus Stevens believed military power was necessary to change the whole fabric of Southern society, "to occupy the heritage of traitors and build up there a land of free men and freedom."[6] Representative George R. Latham of West Virginia spoke in late May of 1866 as the amendment was being debated. He captured the sense of Republican power to establish the South's future and the opportunity it afforded:

> The eyes of the world are on us, and the historian pauses with the ink-dipped pen. What shall he write—that the virtue, intelligence, and patriotism of the American people have triumphed, or that a great people, powerful in war, united by disaster, have failed in the hour of triumph, have proved themselves incapable of securing the blessings and reaping the fruits of victory? Heaven save my country![7]

Idealistic pronouncements aside, Radical Republican Henry Winter Davis of Maryland made it clear that he understood the amendment's larger strategic purpose. "It is a question of power, not of right—a question of self-preservation, not of morals."[8] Democratic Representative Benjamin G. Harris of Maryland, a defiant Confederate sympathizer during the war, provided a reply to Davis's remarks in a speech in the House of Representatives: "You are bent on schemes which seem to contain nothing but the elements of revenge." He perceptively noted that the Republican strategic goal of future peace and security hinged entirely on the success of the freedmen as political actors.[9]

The 1866 Congressional elections were a triumph for the Republicans, giving them sufficient power to marginalize the president while having a dominant majority in both houses to pass legislation without interference from either Democrats or the executive branch. The Republicans were free to execute their strategy for Reconstruction and could take heart from other voices raised in support of the strategy. William G. Brownlow, who prior to the war had been a controversial newspaper editor, was now the provisional governor of Tennessee under President Johnson's reconstruction plan. A fervent Unionist who demonstrated a deep hatred for the Confederacy, Brownlow's newspaper, the *Knoxville Whig*, had lost nothing of the vitriol of its founder. "Taking the ballot from a rebel has the effect of extracting the *sting* from a bee. It renders him powerless for mischief."[10] A similar but far more strident view was presented by the Iowa Supreme Court justice John E. Dillon, speaking at Davenport, Iowa, about the same time. The Southern states should have no representation in Congress, he declared, "until the right of Negro suffrage was granted." Congress should insist on this condition, he continued, and if blacks "fail to get this privilege by law, they would be justified in obtaining it at the point of the bayonet."[11]

The Lincoln-Johnson Reconstruction plan was intended for Southerners to reintegrate as rapidly and easily as possible within the restored national framework. In following the Presidential Reconstruction plan, the former Confederate states sought to establish a degree of autonomy that was far less than the outright independence they had fought for. Inherent in the South's return to the Union was a minority nationalism reflecting a level of dignity and self-respect that Southerners expected the victorious North to acknowledge, as it was essential to their sense of belonging and security. When Southerners faced what they considered humiliation and indignity in the face of a hostile Congress that labeled them as traitors and rebels, a number of warning signs emerged indicating that Southerners would have a hostile response to these threats to their sense of self-worth.[12]

Missouri Republican Francis P. Blair Jr., who would soon break with his party over the Congressional Reconstruction strategy, made public a letter he wrote that was widely printed, criticizing Republicans for disfranchising whites in order to retain their power and position by imposing "measures of unnecessary harshness and indignity in order to irritate and goad our vanquished opponents into acts of resistance."[13] The Charlotte (North Carolina) *Western Democrat* echoed Blair's warning more explicitly: "While North Carolina would do anything that is fair and honorable to restore the Union of the States and promote good feelings among the people, *she will never voluntarily consent to proscribe or degrade nine-tenths of her own people*."[14]

As Congress began articulating its plan of action, a South Carolina newspaper first presented the justification for resistance:

> When the people of the South failed to achieve their national independence for which they battled for four years, their failure entailed no disgrace, neither does our renewed fealty require that we should sacrifice a single feeling or sentiment of manliness. . . . We are the victims of monstrous and most ungenerous outrages, and the revolutionists, in showing their malignant detestation of us, are trampling on the Constitution to effect [sic] our degradation and ruin.[15]

Further indications of the potential danger of a mobilized identity raised concerns from a most unusual source. At a Union Soldier's and Sailor's convention held in Cleveland, Ohio, in September of 1866, the members fully endorsed the Republican vision of "justice, faith in true democracy, vigor, wealth, [and] industry" that would guide the future of the United States. At the same time, however, they cautioned that the vision could not be brought about by force against the South, especially in terms of disfranchising former Confederates. Under the Republican strategy, the Southern states would return to the Union, "but less inclined to patriotism and good faith." Indeed, having taken the measure of Southerners on many battlefronts, the veterans passed on a warning: despite "the error of the cause," the men of the South possessed "personal honor and self-respect . . . will not permit the communities in which they dwell to violate promises given in the most solemn forms."[16]

From Jackson, Mississippi, the *Daily Clarion* decried the use of a constitutional amendment as an instrument of power: "They perceive that there is no other way to get control of the suffrage and introduce the Negroes to the ballot box; no other way to carry such amendments as they wish to engraft upon the Constitution; no other way to subject the domestic concerns of the South to federal control." The *Charleston Daily News* took this protest further, predicting what the action of Congress actually signified. Congressional policy "continues the country in what is actually a state of war and revolution."[17] The *Staunton Spectator and General Advertiser* reprinted commentary from the *Richmond Times* that captured a sense of growing bitterness and anger, resulting from "cruel and tyrannical legislation, remorseless party spirit and political persecution," which the editors concluded, "are accomplishing more to alienate us than the sword would have done."[18]

Senator Charles Sumner of Massachusetts took the lead in establishing the authority of Congress to determine the process for organization and readmission of the states. He presented a series of resolutions on the principles of Reconstruction, declaring the current state governments illegal and that the rebellious states would revert to the status of territories due to their act of secession.[19] These resolutions formed the basis of the Reconstruction Act, passed on March 2, 1867. Its purpose was to establish loyal and more efficient state governments to replace the current governments established under President Johnson that Congress had deemed illegal. The former Confederate states (except Tennessee, which had been recognized by Congress under the leadership of Governor Brownlow, who was instituting a plan very much like that of Congress) were divided into military districts under control of an army officer not below the rank of brigadier general and placed under martial law. The existing governments were considered provisional, and they could be reorganized, restructured, or removed.[20]

As part of the reorganization process from illegal to legal status, each state would call a convention, made up of men who had not been disfranchised for engaging in rebellion, to write a new constitution for the state. This group of eligible participants included individuals who had been declared citizens by the 1866 Civil Rights Act. The only other provision was that individuals selected had been in the state for at least a year. The draft constitution would ensure that all eligible individuals would be enfranchised, and the document would be submitted to the eligible voters of the state for ratification. Once ratified, the new constitution would be reviewed for acceptance by Congress. If approved, the state could form a new government under the constitution, ratify the Fourteenth Amendment, and then elect senators and representatives for admission to Congress.

A supplement to the Reconstruction Act was passed on March 23, 1867, and placed the responsibility for registering voters on the commanders of the military districts, which required an oath qualification for all applicants. After the district commander was satisfied that registration had been completed, he would supervise the process outlined in the Reconstruction Act that would return the state to the Union and have representation in Congress. The great confusion and uncertainty that came about as a result of the Congressional disfranchisement provisions were settled by the military commanders who, in effect, determined voting qualifications of white Southerners through the voting registration process. First, registrars themselves had to be unqualified Unionists; second, every potential voter had to swear what became known as the "ironclad oath," which proscribed white Southerners from voting, not only those who had served in the Confederate army or government, or had held previous offices in the U.S. government before joining the Confederacy (as described in the as yet to be ratified Fourteenth Amendment), but also those in the vague category of having given aid and comfort to the enemy. This, of course, depended on the individual registrar's determination and could include any Southerner not actively or passively resisting the Confederacy's authority. Denying the franchise not only to active participants in the Confederate government and military but also to sympathizers and supporters of the Confederacy was the means to eliminate all potential disloyal elements and to ensure a purely loyal electorate to guarantee a Republican state government capable of implementing the Congressional strategic plan of Reconstruction.[21]

More significantly, it also served to marginalize the Southern-Confederate collective identity that emerged by the end of the war. White Southerners had formed an attachment to the cognitive and emotional features of the Confederacy and had created a cultural conception of themselves as part of a nation, which contributed to an individual, as well as collective, self-definition. By marginalizing white Southerners, the Congressional Reconstruction strategy denied any means for national reunion or restoring loyalty to the United States. Instead, Southerners retreated back to their Southern-Confederate collective identity, creating a condition that would have both unforeseen and significant consequences for the future. Although disfranchisement was not uniformly applied in the states and, in some cases, was not a significant barrier to voting, the threat to the collective identity was far more significant. In effect, the collective identity had been negated and "denied its right as a legitimate player."[22]

A third supplement of the Reconstruction Act, passed on July 19, 1867, gave the district commander specific authority to remove or suspend any officer in the provisional state governments and appoint suitable new officials. District commanders were duty-bound to remove disloyal officials or any officer of the state deemed to be resisting or hindering the military authority. The commander also had complete control of the voter registration boards, appointing and removing members as necessary. Congress emphasized that district commanders had wide latitude to carry out the provisions of the Reconstruction Act and its supplements to achieve as fully as possible its intents and objectives.

To many observers in the North, the Congressional strategy and the approach taken under the Reconstruction Act was clearly defined and understood. Cincinnati Republican Fred Hassaurek, a passionate supporter of the Union cause and editor of *Der Hochwaechter*, an organ of the Freeman Society in Cincinnati, defended the intent of Congress established in the Reconstruction Act. Describing the right to vote as a "power which alone affords security and a fair start in life," Hassaurek asserted that, "the vote is the means to self-sufficiency and the great national attribute of self-protection." He went further, reflecting that enfranchising the freedmen was the path to the "re-establishment of peace and order and security, the revival of business and trade, and the restoration of the Southern States on the basis of loyalty and equal justice to all."[23] The *Delaware Gazette* recognized both the immediate and the long-term outcome of the Congressional Reconstruction strategy: "By enfranchising the blacks, we will not only destroy the Democratic Party and reform our politics, but we will secure the Union. . . . In every Southern State we shall have a strong and unequivocal Union element and in many of these States a large Union majority." As a result, "the country will be nationalized and our people made homogeneous."[24] The *Daily Evening Telegraph* in Philadelphia presented the predominant approach that shaped the Congressional strategy:

> It is a blunder to assume that the principles or details of reconstruction are any longer open to discussion. . . . the policy has been determined. Congress has . . . affirmed its absolute authority over Southern affairs. The time, manner, and conditions of reconstruction have been arranged according to its own conception of duty and interest. . . . What more is needed, except that Congress shall demonstrate its capacity to complete the work it has begun?[25]

Demonstrating this capacity was indeed the issue for Congress. The strategy had been formulated, its concepts for execution outlined, and the structure was now in place to achieve the strategic goals. The Reconstruction Act and its supplements betrayed the attitude of the dominant actor, who overlooks potentially dangerous circumstances in the belief that its overwhelming power would compensate for any emerging threats. The strategy Congress pursued had noble goals, but it had a hard political edge that was based on the firm belief that the people of the South were still enemies and a threat to the nation. This attitude reflected the mood of many people in the victorious states in the aftermath of Lincoln's assassination. Thus, Southerners who had participated in rebellion had to be denied power, and the loyalty of the new governments could only be assured by voters who could take the ironclad oath: unambiguous Southern Unionists, Northerners who had moved South, and the enfranchised freedmen. In executing the strategy, military power was the most convenient and effective means of ensuring compliance, with the added advantage of supervising the implementation of a new voting formula essential to the requirements for transition to readmission. The political leaders of Congress who designed the Reconstruction Act and its supplements confidently based the successful outcome on two assumptions: that the existing military force in the South was sufficient to carry out these tasks, and that the process would be carried out with the tacit approval of white Southerners.[26] Both of these assumptions were far from valid.

The five military district commanders were given a twofold mission under the Reconstruction Act: adequate protection of life and property and the supervision of state governments.[27] The issue of protecting life and property arose from Congress's declaration that the existing state governments were illegal and provisional; therefore, they were unable or unwilling to protect life and property. Supervision of state governments would entail determining loyalty qualifications and establishing a process to secure the political rights of the freedmen. The commanders had the power to suppress insurrection, violence, or disorder and to punish criminals and disturbers of the peace, using either local and state courts or a military court. The provisional state governments were subordinate to U.S. government authority in the person of the military district commander. Supplemental acts indicated that Congress wanted the commanders to exercise their authorities as broadly as necessary to carry out their responsibilities and achieve the goals of the Congressional strategy. But in serving as the way to carry out the strategy, the military commanders were largely detached from the strategy itself. They were following orders, and while given specific requirements intended to achieve the strategic goals, they had little other guidance except to exercise their powers with wide latitude. Missing in the strategic formula, however, were the resources necessary to achieve the goals.

The district commanders selected were all veterans of several campaigns and had commanded at either the army or the corps level. They were combat soldiers with varying levels of success at command. However, none of them had ever encountered what they were about to undertake. While attempting to assess what resources he actually had to establish a level of control to protect lives and property in his district, the commander also had the daunting challenge of carrying out a complete and total re-creation of civil government, which entailed traversing completely new social, political, and psychological terrain. The status of the current provisional state officers was uncertain. Although the commanders had the power under martial law to remove and replace anyone currently in office, the question of how and to what extent they would exercise this power was undetermined and left to the individual commander. To complicate matters further, the president, who was the constitutional commander in chief, was violently opposed to Congressional policy. At the same time, however, General-in-Chief Ulysses S. Grant, who under the Reconstruction Act was Congress's authorized military representative and issued the orders to the district commanders, was disposed to favor the policy. This constitutional power struggle undoubtedly had an effect on how these commanders approached their duties. Generals Philip H. Sheridan, Dan Sickles, and John Pope, who supported the Congressional measures, exercised their authority broadly. Others, such as John Schofield and E. O. C. Ord, were more circumspect.

For the district commanders, the availability of troops was their first concern. With the end of armed hostilities and the soldiers of the surrendered Confederate armies paroled, the Union army began a rapid demobilization. Regiments whose volunteers still had a year or more left on their enlistments were given the tasks of occupation, border observation, and frontier patrolling. Most of the regiments that remained were U.S. Colored Troops with more than 122,000 men representing about 13 percent of the army at the end of the war. Within a year, the number of blacks in the army had already declined to little more than 14,000 men. In late 1867, Union forces in the former Confederacy numbered slightly more than 19,000 men in 10 states at 134 posts and represented the symbolic power that a military district commander could call upon to enforce his will. But that force, in reality, was limited. Only the cities of Richmond, Virginia, and New Orleans, Louisiana, had a garrison of 1,000 men; no other garrison in the South had more than 500 men. Texas had the largest number of soldiers, not as an occupation force, but more of a deterrent to Mexico. In reality, a commander could only call on detachments of 10–15 men; assembling a larger force took a great deal of time.

Clearly the troops available would not make a significant difference and were heavily dependent upon the voluntary compliance of Southerners to be effective. But gaining compliance was sometimes quite difficult as black soldiers created fear, resentment, and hostility among the population, representing as they did the new status of the freedman in the South and a reminder of military defeat. In addition, white police and white soldiers often had violent clashes with black soldiers and demobilized black veterans. In some cases, military commanders requested black soldiers be replaced with white soldiers to avoid exacerbating what Gen. George H. Thomas, commander of federal troops in Georgia, described as Southern "irritation and demoralization."[28]

The district commanders thus had an extraordinary situation on their hands. They were bombarded daily with requests and recommendations for appointments to local government positions, questions on voter registration procedures, requests for legal adjudication, and complaints against local authorities, while also responding to reports of riots, murders, and other disorders. There was insufficient military force to achieve the first part of their orders—protection of life and property—without relying very heavily on the ability of the provisional state governments to maintain law and order. Thus, wide latitude to remove law enforcement officers, judges, and other essential personnel was self-defeating and time-consuming. For every individual removed, the commander was responsible for finding a loyal (ironclad oath swearing) individual as a replacement. Some sort of partnership with the provisional government had to be established if civil order was to be maintained at all. Second, the commanders were responsible for forming entirely new governments, supervising registration of voters who met the stringent requirements imposed by Congress, overseeing the formation of constitutional conventions, and ensuring that voting for ratification of the new constitutions went smoothly and that all properly registered and certified voters could participate.

Under the law, the generals were mainly responsible for creating and maintaining voter registration boards, but establishing the boards and supervising the registration process was far more complicated than anticipated. The generals found themselves spending more time reviewing applications and recommendations for appointments to state vacancies, answering requests for replacement of civil officers, and responding to inquiries regarding the registration of freedmen to vote and requests for assistance in legal matters considered to be unjustly handled by local authorities. The commanders had neither the staff nor the background to address these issues, but nonetheless they had to make the effort because there was no other authority responsible until the election of new state and local governments. Most significantly, the Reconstruction Act and its supplements made no effort to rebuild any relationships among social, political, or economic interests that had been shattered so completely in the South by the war.

The war had opened political, regional, racial, and class cleavages in Southern society that already served as a source of bitter conflict and violence in many parts of the South both during and immediately after the war. Unionists in north Alabama, east Tennessee, northwest Arkansas, and northwest Texas had resisted Confederate control in a mixture of conventional and guerrilla warfare. During the war and in the uncertain aftermath, parts of Missouri, eastern North Carolina, Georgia, Alabama, Kentucky, and Louisiana experienced regionalized and decentralized violence at the community level that included murder, arson, torture, intimidation, and robbery. Confederate deserters—most often pro-Unionist conscripts who had fled the army, draft dodgers, runaway slaves, criminals, and lawless regulators—took control of areas where local or state governments could not enforce law and order. For example, in North Carolina, the Lowry band, consisting of white and black desperadoes, occupied more than eight square miles of territory in Robeson County and battled Confederate authorities while assisting escaped Union prisoners. In these contested areas throughout the South, constant skirmishing occurred as Confederate forces sought to regain control, suppress Unionists, and collect shirkers and stragglers. In Unionist strongholds, Confederate guerrilla chiefs waged their own brand of internecine warfare even after the war ended. The fighting was often vicious. The ambush was the most favored tactic of these armed groups, and raids on stronghold areas were common. Prisoners were tortured and hanged, homes were ransacked, and families murdered in a cycle of revenge killings.[29]

The postwar situation was complicated by the fact that many localities in the South still seethed with the violent antagonisms left over from the war. The cultural and social chaos in the wake of defeat and Confederate government breakdown changed individual and collective attitudes about the role of violence in defending interests, responding to challenges, or solving problems. Because the population had long been inured to war's suffering and brutality, a climate of violence quickly dominated the postwar South.[30]

Between 1865 and 1867, the combination of emancipated blacks who were still somewhat bewildered by the idea of freedom, refugees with no place to go, and returning paroled Confederate soldiers seeking to restore home and family, created social turmoil. The marginally functioning state governments contributed to a sense of chaos, and an economy in ruins required establishing an unfamiliar and uneasy relationship between employer and employee, as each came to terms with the idea of free labor.[31]

The officers of the Freedmen's Bureau, understaffed and overwhelmed by the postwar conditions, attempted reforms at the local level, establishing fair labor contracts, defining new social norms, adjudicating disputes, and promoting schools for the freedmen. These activities were often carried out by single agents whose efforts were greatly limited by public hostility and recalcitrance. Out of necessity, these agents relied heavily on civil and military authorities for assistance. The military commanders who administered Johnson's Reconstruction policy and supervised the Freedmen's Bureau generally avoided interfering with the state governments. Outside of supporting and assisting federal officers in performing their duties, protecting the rights of citizens, and maintaining a force in readiness to deal with a total breakdown of law and order, the occupation forces were careful to avoid issues that placed the military in the role of enforcer. Interactions with the Freedmen's Bureau oftentimes created uncomfortable situations for commanders, who found themselves as the middlemen in disputes and arrests, as they transferred cases from military control to special courts operated by the Freedmen's Bureau, or to civil authorities. There was also controversy as commanders used military commissions under their wartime martial law authority to try cases in circumstances where local courts were either inactive or unwilling to perform their functions.[32] Assaults, shootings, beatings, verbal threats, maltreatment, and killings were commonplace activities; those arrested, tried, and found guilty by local authorities were often whipped, which was the accepted punishment for a wide variety of offenses.

These conditions did not bode well for the Congressional Reconstruction plan. The state elections were intended to be the cornerstone from which new institutions would be built to transform the South. The strategy was based on a strong faith in institutional democracy, individual rights, and freedom that led the Congress to require elections for new loyal governments under military supervision. The expectation was that the sooner the elections for these legitimate governments were held, the sooner the country would be reunited, peace and stability would be restored, and the cost and burden of military occupation would be lifted. In mandating elections in the postwar South, Congress unwittingly created the conditions for further resistance and violence. The state constitutions were intended to strengthen rule of law and the consolidation of peace while advancing economic reform and development. Yet the process of democratization by giving the freedmen the vote only succeeded in dividing Southern society. Because no attempt at reconciliation with the former Confederates was made, those disfranchised had no stake in the government and had no interest in contributing to the development of a stable civil society. The new governments were understood to be imposed as a form of revenge and punishment, preventing any trust or belief in the attempts of the government to establish itself as a legitimate authority.[33]

The state governments created under Presidential Reconstruction, combined with military occupation that offered some level of security and order, along with the support of the Freedmen's Bureau providing relief and protection to a destitute population, had achieved a level of stability that ameliorated political tensions in the fragile social conditions that existed in the postwar South. Although imperfect, a level of stability had been established through democratic norms. As the dominant actor, the Republican Congress, in its one-dimensional view of the Southern states, broke this fragile structure. By declaring President Johnson's state governments illegitimate and instituting new elections intended to punish their traitorous enemies, Congress was demonstrating its dominant power. The new governments under Congress's program were intended to establish the foundations for building institutions in line with the Republican strategic goals, but the cost was the political repression of a significant portion of the electorate. Congress's strategic plan required a significant level of long-term control, largely through military power, to bring about the desired reforms.[34]

Major General Pope, commander of the Third Military District (Georgia, Alabama, and Florida) summarized the difficulties the commanders faced in instituting the Congressional program. In a message sent to General Grant on July 24, 1867, which was widely published, Pope reported:

> It is easy under existing circumstances to win the first victory and reconstruct these States under the acts of Congress, but this victory is only the beginning of the contest and unless it be a victory, openly and fairly won and very decisive in the results, it may prove not only fruitless, but absolutely destructive.

He expressed the opinion that reconstruction must come by the people themselves based on what he called "true principles of government," that is, freedom of speech and press, education, equality before the law, and equality of political rights and privileges. This approach, he believed, would make the process of reconstruction "satisfactory and permanent." If not, "it may become a question whether reconstruction on any reasonable terms is possible."[35] Pope did not have any plan to support his process of reconstruction, but he seemed to understand that some level of reconciliation through judicial and political means was necessary to accomplish social peace and restore normal relations with the former Confederate states.

Under the existing political climate, such an approach that Pope advocated was near impossible. Congress had overturned the existing democratic norms of Presidential Reconstruction with entirely new norms that represented a social and political order in which the existing democratic institutions had to be fundamentally changed to bring about the result Congress intended. The ultimate success of the state governments established through this process depended on the acceptance of their legitimacy by a powerful disfranchised element lacking any political rights. The Reconstruction governments of Georgia, Florida, North Carolina, South Carolina, and Texas were lenient on disfranchisement. Two Reconstruction governments went to greater lengths than others to follow the Congressional intent of disfranchisement of former Confederates. Alabama disfranchised anyone who had also been disfranchised under the Reconstruction Act, copied the restrictions in the Fourteenth Amendment, and added a general restriction on anyone whose actions during the war were deemed beyond the pale. Louisiana took a broader approach by not only disfranchising former Confederate civil and military officers but also anyone deemed an enemy of the United States. These included anyone who voted in favor of secession at the state convention or signed the ordinance of secession, anyone operating outside of recognized military command during the war, and anyone who had publicly advocated treason.[36]

The elections that brought about the new governments appeared to have no intent toward reconciliation, nor was there any sense that future elections would change the existing situation. There were significant problems here. First, the disfranchised whites, who had up to this point held power and influence for generations, were now completely excised from any political rights. In addition, they had fought for the Confederacy or served in the Confederate government and had sacrificed for and served faithfully a cause they believed in. Now they were seeing the true price of defeat: former slaves, who had previously been under the complete domination of whites, and Southern Unionists, who had been considered traitors to the cause during the war, had formed a coalition and were holding political power. Second, what appeared to Congress to be the establishment of freedom, equality, and equal rights appeared to the disfranchised majority as an effort of the North to punish its enemies by imposing an intolerable new political and social order on them.

As property, slaves stood outside the moral order of political citizenship; they had no rights or protections. They were not considered part of civil society. Moving from the status of slave to freedmen under the Thirteenth Amendment made no reference to citizenship, which was a sacred boundary of white Southern identity. When the Reconstruction governments ratified the Fourteenth Amendment as a requirement to readmission, the rights of citizenship were then extended to the freedmen. Southern whites refused to accept the freedmen as citizens or to share the rights of citizenship with them. This threat to Southern identity became a significant factor in mobilization.[37]

As a result, white Southerners had no interest in accommodating freedmen as citizens in terms of equality or suffrage, and they resented the white Unionists, both native and newly arrived, whom they perceived were reaping political advantages and benefits. Thus, neither the state constitutions, the ratification of those constitutions, nor the elections of the governments representing the new order ever had a chance to be accepted as legitimate. The Republican-sponsored state governments were immediately identified with everything the white majority had feared and resisted for nearly a decade. The political viability of these governments was already in question from the moment they were formed. This condition opened the opportunity for the weaker actor to move to asymmetric means to undermine the government's power and authority by subverting the democratic process and employing armed force and violence.[38]

David M. Carter was an officer in the Confederate army until 1864, when he resigned his commission and served in the North Carolina legislature until the end of the war. Under Johnson's amnesty policy, he received a pardon and served in the North Carolina state government organized under the president's Reconstruction policy. Initially siding with Unionists, he turned away when Congress passed the Reconstruction Act in 1867. He made public the reasons for his opposition. "We had surrendered without terms, and there were no precedents in our history to indicate our treatment," he declared. Southerners were now "a subjugated people, entirely at the mercy of our conquerors." Presidential Reconstruction, he believed, failed because the North still believed that the South was disloyal and "sought admission into the Union with a hostile intent." The practical effect of Congressional Reconstruction was to "divest us of all control over the settlement of our future relations with the Union." Establishing universal suffrage for the freedmen and denying it to most of the white population placed the South "forever under the absolute sway of emancipated slaves." The new state governments could be restored by peaceful means with the loyalty and support of white Southerners; without this support, Carter warned, these governments could only be imposed on them by force.[39]

Southern reaction reflected the threat to the legitimacy of the new governments and the dangers that were on the horizon. "The government will be under the direction of the least interested in, and the least capable of its wise administration," a North Carolina newspaper observed, and followed with an exasperated query: "For what? To reconstruct the Union?" The real answer was clear: "for the ignoble purpose of protecting party interests and insuring party supremacy, and the people of the Southern States placed under military despotisms."[40] A South Carolina newspaper saw the obvious result of the Reconstruction Act. "We have said, and we repeat, that we desire peace; but the policy now proposed cannot give us peace." The Reconstruction Act, the article predicted, would only breed strife and violence, which would spread throughout the entire South, reaping a "harvest of crime and blood."[41] Another South Carolina newspaper assessed the political ramifications of the Reconstruction Act, which "invests the negro with absolute political power in each of the ten Southern States, and at the same time, invests him with the balance of power in the United States." The Reconstruction Act, the article noted darkly, "may, for a time, hold us in subjugation to a *quasi*-civil government, backed by military force, but it can do no more."[42] The Richmond *Whig* made the following declaration to Congress: "They have aroused a storm of indignation, which their bloated pride and arrogant self-sufficiency, they did not foresee."[43]

The Reconstruction Act represented a threat to the collective identity of white Southerners and became the basis for the onset of asymmetric warfare in the South. The Radicals considered white Southern resistance to black suffrage and equality as both irrational and immoral. They believed that the force of law was sufficient in power and authority to lead to the improvement of the freedmen through education, example, and toleration. But this rational and moral approach made no sense to Southerners because they understood rationality and morality in entirely different terms within a system of values that defined their collective identity. The imposition of a new set of standards based on the rationality and morality of the Radicals intended to establish a new set of values and a new identity that was a direct threat to the Southern-Confederate collective identity. Southerners rejected completely the system of beliefs of the Radicals and their view of human well-being based on their definition of rationality and morality. They chose to resist rather than conform, to preserve the rationality and morality of their value system on which their collective identity was based.[44]

As the dominant actor, Congress had the opportunity and power to institute its strategy to bring about an entirely new South. Assuming that white Southerners (as the weaker actor) would acquiesce to federal authority and that guaranteeing citizenship and voting rights to the freedmen would ensure their security, Congress confidently relied on the military to organize and oversee the process of restoration. What Congress did not appreciate was that the South was still a region in violent turmoil and that military force in the South was more symbolic than real, unintentionally conveying a lack of resolve to establish the political solution Congress sought. Asymmetric warfare in the South would grow out of this disparity between military means and political ends. Although lacking in power, the weaker actor had an advantage in terms of an understanding of the collective and personal stakes involved in the triumph of the new state governments established under Congress. An asymmetry of interest was already developing that would expose the new governments to some level of resistance. For many white Southerners, a war mentality remained after Appomattox. While Southern independence was no longer possible, a strong Confederate national identity remained alive. It would coalesce with a Southern collective identity as a prelude to mobilization that would lead to asymmetric warfare. In a short time, both state and federal authorities would be faced with entirely new and unforeseen challenges that would take them completely by surprise.[45]

4

Asymmetric Warfare Phase II: The Mobilization of Collective Identity (1867–1868)

"A just God will interpose a shield and buckler for the deeply wronged white men of the South, and under His good providence we have no fear of the result."

The *Daily Phoenix*, Columbia, South Carolina, June 13, 1869

In his December 1868 message to Congress, President Andrew Johnson provided an assessment of what he described as the "party passion and sectional prejudice" that had created the dangerous conditions arising from the Congressional strategy, and he issued a final warning:

> . . . the attempt to place the white population under the domination of personas of color in the South has impaired, if not destroyed, the kindly relations that had previously existed between them, and mutual distrust has engendered a feeling of animosity which, leading in some instances to collision and bloodshed, has prevented that co-operation between the two races so essential to the success of individual enterprise in the Southern states.[1]

Johnson would soon find himself under siege from the legislative branch as rumors arose during the standoff that the president would depose Congress by force of arms, along with growing fears of a renewal of resistance in the South over the establishment of the Republican Reconstruction state governments.[2]

These fears of renewed resistance were entirely justified, but they came to pass in a completely unexpected way. To understand why, it is necessary to appreciate how collective identity plays a central role in shaping the character of asymmetric warfare. The process of collective identity formation and mobilization in the American South provides an explanation for the forms and levels of violence and the political agitation that defines the second phase of asymmetric warfare. White Southerners as individuals identified themselves as part of a collective movement and maintained a cohesion and a commitment that represented the asymmetry of will that neither Republican state governments nor federal authorities could match or overcome.

Collective identity is a "shared definition of a group that derives from members' common interests, experiences and solidarity."[3] This process of self-definition links the personal component to larger social and psychological components that support collective action. As individuals define what their personal identity signifies, a continuum of reciprocal interactions often occurs with others who share the same emotions, values, meanings, memories, and experiences. As individuals establish attachments and commonalities with other people (emotions, sentiments, social norms, conventions, and habits), they begin to act in concert within the group they now identify with. This sense of uniformity and concord with others is essential to collective identity formation and mobilization. Within this process of defining commonalities is the simultaneous process of defining outsiders, the opposition group, as the enemy. Interactions with the opposition group shapes, builds, and solidifies the collective identity, which serves as the catalyst for action while also enhancing group identification and belonging.[4]

Mobilization occurs when individuals identifying with the collective identity take action on behalf of the group. The threshold to action is crossed when perceptions or interpretations of events are believed to be threats to the collective identity. With this mobilization comes a reinforcement of identity and a stronger sense of belonging, which induces a belief that both individual and collective actions will make a difference in shaping conditions, and that such actions will bring about a successful outcome. Most importantly, this mobilization establishes a collectively shared understanding about ends and means. It is this shared understanding that becomes the basis of asymmetric warfare, which manifests itself as issues, tactics, agendas, and objectives aimed at the dominant actor's weaknesses and vulnerabilities, while avoiding its strengths.[5]

Whereas mobilization is essential to asymmetric warfare, collective identity is essential to mobilization. Mobilization is a collective action of a group applying its resources to defend its shared values and beliefs in response to shared fears of a common threat. Collective identity is the means by which resistance is created, formed, organized, and sustained. It is the source for both the decision for action and the form of action, and it defines the desired outcomes. Collective identity carries with it a sense of fealty, representing "an individual's cognitive, moral, and emotional connection with a broader community as a perception of a shared status and belonging, which shapes an individual's sense of self and personal identity."[6]

Social, political, or economic networks often play an important role in developing and sustaining collective identity. Collective identity is expressed in various cultural forms, such as names or titles, narratives, symbols, rituals, clothing, or identification by race and nation that serve to confirm common interests and common bonds. Through these common interests and bonds, individuals are committed to contribute overtly, covertly, or passively to collective action. The imperative for collective action emerges from already-established interpersonal networks within communities, where the importance of word of mouth, trusted relationships among neighbors, local interpersonal relationships, and kinship are crucial to success.[7]

Identity mobilization involves both structural and cultural factors, which made the white South more prone to resistance and violence. The antebellum hierarchy of Southern society had been transformed during the war. Soldiers and civilians of all classes had endured the trial of war and had conducted themselves honorably. Defeat had not brought shame or repentance over the issues that had led to the war, nor had surrender brought disgrace to the people's heroic and patriotic demonstrations of devotion to the nation. Southern society was disposed to violence as part of the cultural factor of personal honor that accepted a certain level of violence as normal. The death and destruction of war only made violence more normative.

Identity mobilization welds individuals to a group, creating a sense of belonging and common purpose directed toward specific goals and objectives, which are defined and understood through a process of framing. Frames are interpretive packages that serve to inspire and legitimate all levels of violent and nonviolent action by identifying the events and experiences that have created the conditions requiring collective action. Frames define both the nature of the cause and the threat to the collective identity. From this framing process emerges the strategic and tactical approaches of the asymmetric actor.[8]

The Confederacy had been created to revalidate the essential principles that had served as the basis for the very nature and purpose of the United States. Loyalty to the Confederacy demonstrated the Southern faith in the essential values contained in the Constitution. Confederate national identity was tempered in the crucible of a total war for independence that superseded class status in shared sacrifice and hardship. Within this collective identity were two frames: Southern virtue and honor composed the frame of cultural superiority; patriotic duty, sacrifice, and courage were the components of the frame of stoic endurance. The Southern identity frame included the self-image of honor, which entailed the obligation to use violence to settle scores where personal honor was involved. This was combined with the experience of battle in which Southerners had demonstrated a stubborn courage and selfless patriotism in the individual soldier's conduct on the battlefield and the respect his enemies paid to him. The South's social institutions, established by cultural norms and social hierarchy, served as the basis for a wartime collective identity. Southerners endured as people who were uniquely different from the Yankee enemy seeking their destruction. They attributed meaning and significance to the suffering they experienced as a people and as a nation; it had served to reinforce their faith in eventual victory and the worthiness of their cause to gain independence. However, when victory was denied, they accepted God's inscrutable purpose with fortitude and humility. This sense of virtue and superiority was at the heart of Confederate national identity—Southern men lived up to the martial ideal of selfless courage, and Southern women demonstrated a noble forbearance. White Southerners emerged from the war with a Southern-Confederate identity that contained a collective self-description and self-understanding that provided a sense of commonality and connectedness.[9]

Although the Confederacy had collapsed with shocking rapidity, Southerners were not altogether willing to give up frames of the Southern-Confederate collective identity that set them apart from the North. They remained intact after the war, providing the white Southern mind-set with a powerful sense of distinctiveness and separateness, along with a unique combination of defiance and accommodation. Southern civilization was superior and remained so, even in defeat. Defeat, in fact, only continued this process of fusion and sense of solidarity. The Confederate armies had been feared and respected to the very last days of the war. The famous salute of the Army of the Potomac to the Army of Northern Virginia at Appomattox made that respect clear. In the aftermath of the war, final victory or defeat was less important to the former Confederates than a recognition of honorable conduct and duty faithfully performed.[10]

Under Presidential Reconstruction, Southerners were willing to accept reunion based on loyalty to the United States as well as accepting the conditions the war had imposed upon them, defined by the existence of newly freed slaves. In return, Northerners would accept a modified status quo ante bellum marked by white Southern local control, in which the South would determine its own place in the new Republican order. Reconstruction and reconciliation would thus be marked by a mutual demonstration of honor and respect that would serve to temper the political discourse of the postwar order. The Congressional response was to forgo any attempt at reconciliation and to punish rebellion and treason, while undertaking a strategy to remake the South with the goal of creating a region that would ascribe to the Republican ideals of racial equality and material and moral progress. By rejecting the idea that the secessionist South could be loyal, and by denying both respect and honor to their defeated countrymen, the Republican Congress was making a direct attack on the Southern-Confederate collective identity. The intent of establishing state governments sustained by the votes of newly enfranchised freedmen only served to heighten the sense of Southern unity and also sharpened the factor of white racial identity, which, in response to this threat, would seek to ensure that the South would belong, both in the present and in the future, to the white man. Thus, the transformation of the freedmen into citizens and voters as the centerpiece of the Congressional Reconstruction strategy created social, economic, and political threats to the collective identity that served as the basis for mobilization.[11]

The new postwar political and social order created high levels of fear and anger and defined the threats to the Southern-Confederate collective identity. In response, white Southerners in localities began uniting through common interests and bonds, feelings of group membership, and shared grievances related to demands for rights and privileges. In doing so, the collective identity created and defined the boundaries between friends and enemies and established a set of commonly held requirements for peace and security. Thus, the Southern-Confederate collective identity defined the enemy as the carpetbagger interloper from the North, his white Southern scalawag ally, and the dangerously radicalized freedmen. The state governments under Republican control wielded power through agitation and proselytizing among the freedpeople and through the formation of Union Leagues and militia units to suppress white Southerners. Peace and security would come when white Southern claims to status and recognition were achieved through the reassertion of traditional political control and the neutralization of the freedman as a political actor.

The white South's social order supported its capability to mount an effective asymmetric resistance. But without a common identity, individuals cannot form as a collective agent, creating a continuity for action that is necessary for mobilization. The Southern-Confederate collective identity created a sense of unshakable solidarity that could be mobilized to accomplish a shared purpose. In the context of asymmetric warfare, the mobilized weaker actor faced a dominant actor that appeared to present a daunting united political front. However, the ability of the dominant actor to maintain its unity of purpose in the North was not assured. In addition, the newly created state governments were vulnerable. For the resistance to succeed, the weaker actor had to maintain a limited invulnerability to the dominant actor's power, while at the same time maintaining a steady level of violence and disorder against the stability and legitimacy of the state governments in order to impose increasingly higher material, political, and psychological costs on the dominant actor.[12]

As the weaker actor, the mobilized Southern-Confederate collective identity would initiate asymmetric warfare to control the information battlefield to challenge the Republican ideology of Reconstruction and weaken the will of the dominant actor. In order to weaken the source of power of the state governments established under the Congressional Reconstruction strategy, the resistance would employ various forms of violence, or threats of violence against the freedmen and their white allies. Political agitation, backed by the threat of, or actual use of, violence would also be used to undermine the legitimacy of the state governments.

After the Southern-Confederate collective identity had been mobilized, a threshold had been crossed. Equilibrium and reconciliation were no longer possible without violence. Asymmetric warfare presented opportunities for the weaker actor to begin a campaign of armed violence directed at the identified enemies of the collective identity. As a reflection of identity, this violence would serve multiple purposes and create conditions that represented a direct challenge to the Congressional Reconstruction strategy.

Secession and war had created a distinct Confederate national identity reflecting the deep cultural convictions of white Southerners. The Confederacy was a nation in arms, sustained by the ideology of Southern independence, whose transcendent purpose had become part of white Southern culture. As a people, they had fought against an alien entity hostile to their very existence and had failed. The war effort represented the importance of sustaining that national identity, and the meaning the people attached to belonging to that nation had not disappeared with defeat. White Southerners constructed a collective memory in the wake of defeat reflecting a sense of Confederate virtue that represented courage, duty, and a deep sense of honor. It is significant that women of the South began the practice of decorating the graves of the Confederate dead at the same time a Southern-Confederate collective identity was being formed in response to Reconstruction. This act of remembrance was a powerful symbol of identity and loyalty to the Confederacy that helped individuals coalesce together around shared beliefs, ideals, and memories. White Southerners in the emerging collective identity viewed themselves as having acted in good faith after the war and submitted to the prescripts of Presidential Reconstruction; but the North had called them traitors and imposed despotic regimes dominated by invaders who brought with them their "Puritan fanaticism, miscegenation, and corruption" to oppress the South.[13]

The threat to the collective identity began to take form as Southern newspaper editors invented the carpetbagger and the scalawag, powerful symbols essential to the weaker actor in asymmetric warfare. Southerners used these terms as "imaginative anchors" that represented a larger set of presumptions, explanations, and prescriptions essential to the mobilization and sustainment of the collective identity.[14]

This mobilization threshold, marking the second phase of asymmetric warfare, occurred shortly after the Republican Congress imposed its program of Reconstruction on the former Confederate states. Politically marginalized white Southerners were confronted with a stark choice of either submitting to the newly established Republican state governments or resisting them through violence. Armed assaults, terrorism, and violent reprisals were more attractive to the weaker actor than accepting the new conditions imposed by Congress under the Reconstruction Act largely because Southerners believed participating in violence furthered their interests and supported their desired outcome—a recognition of legitimacy denied through war but obtainable by other means.

Inherent in this phase of asymmetric warfare is violent conflict related to electoral politics, where the dominant actor seeks to achieve its goals by establishing a new set of democratic norms through the election process. The dominant actor generally views elections as ends in themselves, that is, a form of resolution marking peace and stability, giving newly elected officials time to establish the structures and capacities to govern effectively, and implementing the strategic goals and objectives of the dominant actor. Lacking the ability to influence the distribution of power in government, the weaker actor seeks to check the power of the dominant actor in other ways. In asymmetric warfare, fighting becomes a substitute for voting. In the second phase of asymmetric warfare, the level and persistence of violence, regardless of its purpose, assists in advancing the interests and ultimate goals of the weaker actor. Thus, initiating violence has the traverse advantage of increasing the weaker actor's strength, regardless of whether or not the dominant actor chooses to employ force.[15]

Violence in asymmetric warfare serves partially as armed propaganda, intended to manipulate perceptions about the strength and influence of the resistance, define the movement's purpose for action, and raise popular consciousness. Armed action is intended to seize the psychological advantage by displaying an image of strength and credibility to gain broader popular support. As the perception grows that the dominant actor is incapable of dealing with the scope and scale of the resistance, success breeds a broader base of support as people make choices based on their collective identity and associate themselves with the resistance group. As benefits accrue based on the perception of the dominant actor's weakness and the resistance's strength, the level of support and recruitment grows, and the resistance becomes self-sustaining. Both individuals and groups are induced to take action in the belief that their efforts are an essential part of the overall resistance. Along with this impulse to action is a clear appreciation of both the limitations and the advantages afforded by the environment. Particular courses of action emerge that assist the weaker actor in pursuing its goals in the face of the powerful forces arrayed against them.

Asymmetric warfare is both politically and privately driven. Armed groups initially have the goal of delegitimizing the state, but as success and the incentives to participate grow, criminal elements may also join, taking advantage of the resistance's popularity and the weakness of the state government to enforce law and order. Individuals also take care of personal vendettas under the guise of the resistance. Violence at the local level relates to private quarrels, revenge, or vigilantism—punishing transgressors for minor crimes such as theft or enforcing worker discipline. Individuals and the criminal element take advantage of conditions to settle old scores or simply to engage in violence for its own sake. These actions have no relation to political issues but appear to be a concerted and coordinated action at the aggregate level.

In the Southern states, race riots, night riding, political agitation, and newspaper articles promoted ferment, which served as an alternative means of both maintaining social control and altering the level of political support for the Republican state government. These forms of direct violence were effective because of the support and collaboration of the majority of individuals in the community. The motives for the violence were not always clear—sometimes personal and sometimes abstract by intent of the armed resistance—but the same kind of violence was employed by independent actors who used these acts for their own purposes. The mix of ambiguous actors operating at multiple levels of intent, uncontrolled and undirected, contributed unintentionally to the overall effect desired by the active armed resistance—to delegitimize the Republican governments and establish a self-sustaining movement that was beyond the capability of the state and local governments to control.

The Republican governments in the Southern states, made up of transplanted Northerners, Southern loyalists, and free blacks claimed de jure power, but that power rested on a fragile coalition, heavily dependent on the freedman as a block of reliable voters, without whom the probability of winning future elections was slim. White Southerners, however, by initiating a level of violence and demonstrating a mobilized identity, represented de facto power.[16] The violence and disorder presented the image of success and an incentive for others to participate, while at the same time reducing incentives for freedmen and loyalist whites to maintain support for the Republican state governments. The intent was to present a threat, whether real or imagined, to these key supporters so that they would withdraw their support and begin to conform to the interests of the resistance, weakening the state's ability to maintain its security and legitimacy. When combined with nonviolent political activities, such as political agitation and opposition, the state governments would be further pressed to establish legitimacy.

In 1867, as the Congressionally mandated state governments were being organized under military supervision, Southern newspapers raised the specter of a threat to white Southerners that served as a primary means of mobilizing and sustaining identity for resistance. This threat was the arrival of Union Leagues. The Union League was founded in Philadelphia in 1862 to support President Lincoln and the Republican Party. The League not only strongly supported emancipation for the slaves, it also endorsed black enfranchisement as an outcome of the war. In the immediate period after the surrender of Confederate forces, the League expanded into Savannah, Nashville, and Richmond where U.S. troops were on occupation duty. The Union Leagues attracted a cadre of prewar free blacks, black veterans, white schoolteachers who had migrated South to provide education to the freed-people, Freedmen's Bureau agents, black preachers, and mixed race profes-sionals who desired to organize the freedmen and prepare them for exercising their newfound political rights and assist them in achieving the Northern ideal of economic independence through free labor. The Union Leagues, or Loyal Leagues as they were sometimes called, became an arm of the Repub-lican Party in the South and represented a powerful means to sustain the Republican Reconstruction governments and, by extension, to achieve the strategic goals of Congress. The Union Leagues thus served as an instrument for supporting political control by uniting the white loyal minority with the freedmen.[17]

The loyalty the freedmen demonstrated for the Republican Reconstruc-tion governments was largely related to the potential power they believed would come their way and allow them to take advantage of the material and personal benefits they would enjoy under the new postwar order.[18] The Union Leagues were an excellent example of this belief in achieving power through a demonstration of loyalty, but this demonstration was based on the expectation of certain incentives or benefits to elicit political coopera-tion. Because most freedmen were illiterate, they could only be informed and organized through personal communication and by rallies, meetings, and assemblies. The freedmen joined the Union Leagues based on the incen-tives offered by organizers, who guaranteed that the Republicans would make good their promises of security through the ballot and a strong, secure, and stable Republican-controlled government that would guarantee their rights.[19] The *Tri-Weekly Standard*, a pro-Republican organ in Raleigh, North Carolina, stated the goal of the Union Leagues was "to promote the cause of loyalty to the Union." The activities of the Union Leagues served also to impart "a conviction of security and strength," and concluded that the Leagues were the only means to win a glorious victory for the Union in the South. A later edition published a Union League charter application form.[20]

As the activities of the League became known, reports began to circu-late that freedmen, under the active encouragement of their Republicans organizers, were holding secret nighttime meetings; armed freedmen were organizing in a quasi-military fashion and conducting drills; and acts of nighttime arson were directed against white property owners. These reports led Southern newspapers to respond with consternation and alarm and fueled a belief that armed blacks, led by white radicals, would be encouraged to wreak revenge against the former Confederates.

An article in a South Carolina newspaper and published elsewhere reported on the Union League ritual, "copied from the printed book." The object of the Union League was to preserve liberty and perpetuate the Union, "secure the ascendency of American institutions on this continent," and "to protect, defend, and strengthen all loyal men and members of the Union League . . . in all their rights of person or property." It included a pledge of secrecy as well as a pledge "to resist to the utmost extent of your power, all attempts to subvert or overthrow the Government of the United States," which included "insurrection, invasion, or rebellion." The member pledged to do everything within his power to elect "true and reliable Union men." Elaborate initiation ceremonies were held, accompanied by oaths, secret signs, and symbols supported by a political catechism that bonded members to the organization and its cause.[21]

A North Carolina newspaper carried the results of a trial related to the murder of a young white man by white and black members of the Union League in Pickens, South Carolina; a South Carolina correspondent identified by the moniker "Cosmopolie" reported on the rituals and signs associated with the Union League and the Red Strings, another Unionist organization, in North Carolina. A Memphis, Tennessee, newspaper reported that the objective of the Union League forming in Louisiana was to "release Louisiana from rebel rule." In Nashville, Tennessee, another reporter identified only as "Umbra," revealed that the Nashville post of the Grand Army of the Republic was cooperating with the Loyal League, preparing for an attack on "Rebels" and their property. Freedmen at the Loyal League meetings were subjected to a "bloodthirsty and fiendish" address. The freedmen of the Loyal League were to be organized into companies, given arms and ammunition, and trained to "slay without mercy all opposers." Spies and informers of the Loyal League were active in collecting information about potential targets.[22]

An article in the Pulaski, Tennessee, newspaper described the white leaders of the Union Leagues as nothing more than deceivers of the "deluded negro." In doing so, they had forfeited the respect of every honest man and deserved only scorn and contempt, and should be considered "more dangerous than a man who would steal your horse or pick your pocket."[23] One Louisiana newspaper offered a more measured approach. Noting a report that 300 armed freedmen were seen in Rapides Parish in military order, carrying flags, and led by three Radical whites, the newspaper observed that the only way for peace was for the freedmen to "leave off politics, kick every Radical leader, go to work, strive to an honest living and better their condition."[24]

As Southern newspapers fed the fears of the threat posed by the Union Leagues, the first signs of mobilized resistance began in Tennessee. Because Tennessee had been readmitted to the Union, it was not under the Reconstruction Act. As early as April of 1867 and into the summer of that year, acts of low-level resistance emerged. Voter registrars were harassed and threatened; there were numerous incidents of vandalism, and several individuals, black and white, were killed. Governor Brownlow vowed to defeat what he called "the spirit of nullification and rebellion" that still existed in the state. Over 1,000 men of the Tennessee state militia, most of whom were east Tennessee Unionists, were deployed to central Tennessee to protect loyal voters, as the Union League reportedly marched freedmen to the polls in a patriotic parade to vote. The militia detachments faced public confrontations, rock throwing, and taunting.[25]

Major General Thomas commanded 1,000 U.S. troops in Tennessee and sent detachments to Nashville and Memphis to prevent violence, but the soldiers were only to act if there was a confirmed level of disorder that civil authorities could not control.[26] Similar types of violence occurred in Georgia and Louisiana, where any large assembly of blacks was liable to be broken up, ballots destroyed, and voter registration officials, influential freedmen, or loyal whites were targeted and attacked. In some cases, armed groups of blacks and whites battled in the streets.[27] This was a sign of things to come.

By June of 1868, Florida, North Carolina, South Carolina, Louisiana, Alabama, and Mississippi had been readmitted to the Union under the Congressional Reconstruction mandates. With readmission, military oversight was removed, and the new governments were expected to begin functioning under the new constitutions approved by Congress. But the democratic process that Congress had outlined in the Reconstruction Act could not guarantee a democracy. Stability was the first essential for democratic norms to establish themselves. The election of a new government by itself could neither establish legitimacy nor order, nor establish any level of stability necessary for government to function effectively.[28]

As a society in transition, the South was already subject to high levels of violence and social conflict, which by themselves represented a direct threat to building democracy and democratic institutions. Individual acts of violence to settle scores were common and were part of the fabric of life in the South. The Reconstruction governments were financially, organizationally, and politically ill-equipped to address all the challenges they faced. Communities were separated by wilderness, and poor roads hampered travel and communication. Even prior to the establishment of these new governments, laws were enforced haphazardly, depending on the availability of a magistrate of sheriff.

The transition from military oversight of a provisional state government under Presidential Reconstruction to the Congressionally mandated

Reconstruction governments presented a serious challenge to achieving the strategic goals sought by Congress. The new state governments had to rush to fill the vacuum created by the demise of the provisional governments before all power and authority evaporated completely. This was a formidable task. The new state governments replaced the old magistrates who could not swear the ironclad oath with loyal appointees or, more often, through local elections. The states traditionally relied on the local authorities for the legal, commercial, and social order. Local courts formed a crucial link between political and economic power of the new state governments. County magistrates had responsibility for law and order, with very broad arrest power as well as the power to try minor offenses. They were also expected to oversee business and social activities. County courts and county superior courts were responsible for maintaining law and order.[29]

Under the original U.S. Constitution, the states determined who was a citizen of the state and, thus, who possessed civil rights within the state. The Fourteenth Amendment changed that situation, making U.S. citizenship also state citizenship and placing the federal government in oversight over the state and local authorities in maintaining the proper relationship of the citizen to the state by guaranteeing civil rights. Freedmen were now subject to the state laws just as any other citizen. As citizens, freedmen were connected directly to outlets of state and federal power in ways no American citizen had been before. Under Congressional Reconstruction the relationships among local, state, and federal law and governance were reordered. Protection and enforcement of individual rights became more important than social order. Individuals were no longer subordinated to the social order; instead, federal, state, and local law under the Fourteenth Amendment was now actively involved in protecting individual rights.[30]

This new situation put untested local magistrates and law enforcement personnel in a difficult position of guarding and defending rights of individual citizens as opposed to maintaining general peace and security, which had been their primary roles prior to the war. They were especially responsible for protecting the rights of the new citizens created under the amendment who had expressed loyalty and support to the new government. The state governments would themselves have difficulty in carrying out the intent of the amendment. The Mississippi Republicans, for example, recognized that voting rights for blacks was the only insurance that allowed their government to survive. Any type of power sharing agreement with former Confederates, or allowing significant numbers of them to vote, was a formula for political suicide. Because the state governments were heavily dependent on the popular support of the freedmen and Unionist whites, they were hampered almost from the outset by internal instability as a result of factional infighting and the competition for power and influence in the new government structure.[31]

The new state governments were nominally biracial but heavily influenced by Northern whites who now resided in the South and were heavily represented as delegates to the state constitutional conventions or served in a majority of the influential leadership positions. The *Weekly North-Carolina Standard*, a pro-Republican newspaper, made approving remarks concerning the state constitutional convention that were typical in reflecting the intent of the Congressional strategy. It described the "earnestness" of the 150 leading Unionists of the state and the 150 "colored" delegates, meeting on the same floor and "co-operating in the most cordial and harmonious manner" that aroused "emotions of gratitude and patriotism."[32]

A large number of the men who made up the ranks of the delegates in the state constitutional conventions were former Union army officers who had led U.S. Colored Troops units during the war. Blacks were a majority in the South Carolina convention, and blacks had strong representation in Louisiana and Florida as well. Loyal whites were strongest in North Carolina, Arkansas, Alabama, and Georgia. Conservative whites were more numerous in Virginia, Mississippi, and Texas.[33] The state constitutions that were submitted for approval reflected the strategic goals and intent of the Congress. They established a broad democratic government intended to open opportunities to loyal whites and blacks based on individual rights, access to public education, and economic initiatives that would serve to bring about the economic integration with the North's market economy. The Union men from the North served in a number of important positions in these new governments and were therefore responsible for bringing about the transformation of their states as they were returned to the Union. The leaders were attempting to make significant advances with an uneasy and volatile political coalition.

The freedmen and loyal whites had formed what has been described as "a politics of dependence."[34] While Northern Union men and blacks were fully committed to the Republican ideology and were among the most strongly attached to the Congressional intent of Reconstruction, the loyal Southern whites were often far less comfortable with the reformist agenda of the new governments.[35] They accepted the Congressional strategic requirement of enfranchisement of the freedmen as essential for Republican control. Sen. Charles Sumner of Massachusetts, one of the architects of the Congressional strategy for Reconstruction, admitted as much, saying, "Without the colored vote, the white Unionists would have been left in the hands of the rebels; loyal governments could not be organized. The colored vote was a necessity."[36]

Despite their support for enfranchisement in principle, to many loyal whites, establishing a broad set of rights for the freedmen was less important than keeping the former Confederates out of government. Although there were many loyal whites who were strongly committed to the Republican ideology, the scalawags in general were a weak reed upon which to sustain a stable, effective, and legitimate government.[37] Republican Gov. Robert K. Scott of South Carolina captured the essence of the dilemma in his message to the legislature on November 27, 1868:

> A Government unable to enforce its laws and protect its citizens is a mockery and a sham, meriting the scorn and contempt of its opponents and unworthy the confidence and support of its friends. It cannot administer justice through its Courts, or collect its revenue by taxation.[38]

Newspapers throughout the South were a powerful influence in undermining any support for the new Republican governments. A Louisville *Courier* correspondent reporting on the Georgia constitutional convention in September of 1867 reflected the dominant attitude of most Southern whites that the convention would:

> ... purport to represent by savage Africans and white men who can take the test oath. ... The effort is to hunt out men who are either prominently obnoxious by their treachery, ignorance and general want of reputation or who are so utterly obscure and inexperienced that no human beings can say what opinions they hold or what course they will adopt.

A South Carolina newspaper was far more threatening:

> ... it is not a question of negro equality and negro rights that is involved in the Radical Reconstruction, but of white rights and white equality; ... every one of them that votes for this constitution declares himself the enemy of the white race and must be so held and treated in every relation of business and politics.[39]

A Louisiana newspaper reported that the new Republican state government, which could rely on a registered voting majority of more than 38,000 black men, would assure that Congress could rely on Radical representatives in Congress and Radical electoral votes. Louisiana had become a model state for following the Congressional Reconstruction strategy, delivered to the Congress by General Sheridan, "whose sole thought, day and night, was of giving the political power in the State to his political friends by fair or foul means."[40]

The newspapers employed a number of rhetorical weapons against the military district commanders responsible for the implementation of Congressional policy. General Sheridan was the "Duke of Louisiana," and General Sickles was referred to as "the satrap of the Carolinas." Both officers were "gingerbreaded all over with the trappings of vulgar despotism" that was "concealed under the specious disguise of 'progress,' 'reconstruction,' and 'equality without regard to color.'"[41] Despite the trappings of power in the persons of the officers commanding the districts, their actual power to enforce their will was very modest. In 1868, there were about 17,600 U.S. troops in the South; a year later, there were only about 11,000.

Southern newspaper editors fueled mobilization by giving a voice to the collective identity of white Southerners by creating the symbolic representations that defined the enemy: carpetbaggers, scalawags, and their black allies. The carpetbaggers were self-seeking Northern adventurers, the scalawags were the traitorous white betrayers, and blacks were viewed largely as either pawns or ignorant dupes, more or less neutral and subject to whatever influence a group could apply in its favor. With this sense of Confederate virtue and Southern collective identity threatened, mobilization was a natural course of action.[42]

To white Southerners of this mobilized identity, the carpetbagger and the scalawag were the authors of their subordination and also the targets of an ideology of resistance. By 1868, carpetbagger was becoming a common term in Southern newspapers, and was also being used in Northern newspapers. The *New York World*, for example, described the carpetbagger as one "who has no stake in the country beyond his satchel," but the political conditions of Radical Reconstruction allowed him to rule a Southern state. A Jackson, Mississippi newspaper had a far more blunt definition. Carpetbaggers were "white men who are striving to build up an African despotism upon the overthrow of liberties and the social institutions of their own race."[43]

The internal weakness within the state governments was an immediate indication of a lack of legitimacy. This obvious weakness, combined with the dominance of carpetbaggers and freedmen as the main supporters of these governments, only strengthened the growing trend to mobilization of the collective identity of the weaker actor. Furthermore, weak local policing and a government that demonstrated real or perceived ineptness and corruption became the target of a mobilized opposition that combined grievances with an opportunity for action, whether nonviolent political agitation or violence. A Louisiana newspaper succinctly summed up the situation. "Whatever may be the next outrage a Radical Congress shall perpetuate upon the Southern people," the writer intoned, "we advise our friends to stand firm. . . . We now know our strength."[44]

Indeed, the weaker actor had sent innumerable signals that a mobilized collective identity existed and that it was ready to initiate asymmetric warfare. Knowing its own strengths, this mobilized identity was prepared to exploit the stronger actor's vulnerabilities through a course of action that represented a completely unexpected approach.

5

Asymmetric Warfare Phase II:
The States Under Siege (1868–1870)

In the beginning of the second phase of asymmetric warfare, the mobilized resistance lacks sufficient organization and power to change the political status quo. As the weaker actor, it seeks to destabilize the government by delegitimization—to demonstrate the government's inability to execute its basic functions. The battlefield in the second phase comprises the small towns and backwoods where groups can operate securely and clandestinely, spotting and targeting groups and individuals who are assessed to be a threat or capable of ensuring the proper function of government. The cumulative effect of this disordering violence, protest, and agitation would attract state-wide and national attention, all the while highlighting success and immunity to the power of the state authorities in order to build further support and active participation. This early phase marks the emergence of the Ku Klux Klan. (During the Reconstruction period, contemporary observers and newspapers used the term "Ku-Klux" to describe the Klan, so this term is used throughout the book to reflect the way Americans understood events of the time.) Klan activity represents a prominent factor of asymmetric warfare—actions by the weaker actor that are entirely unexpected and unanticipated by the stronger actor.

As the subject of endless study and high moralizing, the Ku-Klux has been described in about every way: terrorists, guerrillas, white supremacists, the paramilitary arm of the Democratic Party, and the instrument of the elite. None of these labels have ever been defined to any degree, nor has the Ku-Klux and its methods been understood in a strategic context as a component of warfare. While often associated with political action, the Klan has been given a monolithic status, and the term is often used as shorthand for any act of violence in the South during the Reconstruction period.

In the strategic context of asymmetric warfare, the Ku-Klux is a representation of identity mobilization, crossing the threshold from shared fears and animosity to direct action. Individuals were acting in concert within a group with which they identified. The Ku-Klux represented role identity—the adaptation of a self-meaning to accompany the role performed. It represented the uniformity of perception and action that came from the mobilized collective identity. The Ku-Klux phenomenon merged group and role identity with individual identity as a spontaneous action, spawning innumerable imitators from state to state.[1]

Klansmen's theatrics were a reflection of role identity—their costumes disguises, rituals, language, and violent, as well as nonviolent, actions represented a Southern-Confederate identity of resistance to federal- and Republican-controlled state authority that challenged the inherent inequity and lack of power the new state constitutions had imposed on white Southerners. Disguising intentions and purposes were intended to create uncertainty in order to distract and delay authorities from taking action. The theatrics were also intended to create a sense of power and success to sustain and increase popular support. As more and more individuals chose to identify themselves as Klansmen and the image of the Ku-Klux took on a life of its own, illustrating the power of the image through costume and performance behavior became the incentive to cross the threshold from passive to active resistance.[2]

The Ku-Klux established the postwar boundaries between those who shared the Southern-Confederate identity and those who did not. The Ku Klux deliberately identified themselves to their targets, and their attacks were intended to establish themselves as society's defenders and guardians from the outsiders who had placed themselves apart through their political acts, their violations of social norms, or petty crimes. Klan groups were locally organized and focused. The Ku-Klux demonstrated community order through their organization and discipline as well as their planning. Even the outlandish outfits and mysterious regalia were intended to demonstrate unity and solidarity.[3]

There is no evidence of centralized control or strategic direction; even at the local level, different groups may have coordinated, but few strayed away from the local area where they knew the land and the people. Thus, it became easy to deny the Ku-Klux existed at all because it was so unorganized and spontaneous. Costumes and disguises allowed all types of behavior. A bored young man, a battle-hardened veteran, a disgruntled farmer, or a criminal and outlaw could don a costume and step outside all limits of morality and convention.[4]

As a reflection of identity, violence was part of making a statement of resistance and illustrating the inability of state or local law enforcement to stop them. The rise of the Union Leagues had a corresponding rise in Ku-Klux formation. Southern newspapers demonstrated a great outcry against the Union Leagues and justified the Ku-Klux as a legitimate response. Although often downplayed by historians, the existence of strong fraternal interest groups heightened by political differentiation such as the Union League and the Ku-Klux guaranteed violence and conflict. Throughout the South, the idea of groups of armed freedmen created uncertainty and extreme security fears among whites, who feared that unless they initiated preemptive violence against the threat, they would be killed first. If freedmen attempted to form armed groups to provide for a higher level of collective security, whites took actions to protect themselves, thereby triggering a spiraling security dilemma that inevitably would lead to a collision.[5] For example, armed white groups responded to the rise of the Union Leagues with raids to seize private firearms from freedmen and, in some cases, to steal weapons and ammunition from the state government destined for the state militia. Many of the most well-known violent events at Camilla, Georgia; Colfax, Louisiana; Vicksburg, Mississippi; and Hamburg, South Carolina had their origins in this security dilemma.

Over time, organized violence became more focused as certain individuals were recognized as political actors. Attacks against these targets generally were well-rehearsed and intended to communicate a specific message to both the victim and the community. Political leaders, individuals of influence among the freedmen, or participants in black political organizations were major targets, which served the dual purpose of preventing these organizations from functioning effectively at the local level and weakening confidence in the Republican Party. Violence was at other times related to accepted standards used for punishment of slaves for infractions such as contract violations or petty theft, or, more significantly, to reimpose a level of control and obedience.[6]

The first public mention of the Ku-Klux appears, not surprisingly, in Pulaski, Tennessee, the purported birthplace of the Klan in the fall of 1867. The newspaper noted the presence of "armed bands of men parading our streets at unusual hours of the night." Mayor W. F. Ballentine, in a letter to the editor, reported a conversation with the Freedmen's Bureau agent, who complained that the Union League meetings were being disrupted by the Klan, who he stated were always armed. The mayor reported that he had no knowledge of the "KuKlux," had only seen them once or twice, and did not know the object of their society, but asserted that they had not disturbed anyone. The mayor made it clear in his letter that any armed body of men parading the streets would be arrested and turned over to the state authorities, "if it took every man in town to do it."[7]

A few months later, there were published reports of a rising level of fear among the freedmen in western Tennessee with the emergence of a secret order in Maury and Giles counties called the "Ku-Klux Klan." No one seemed to know what the group's purpose was or its membership. The article noted that the group maintained an "unbroken silence" while on parade and observed that "they dress in long red gowns, red pants, and red caps, with black face cloths covering their features." Although Radicals dismissed them as nothing more than "Rebel bush-whackers," no one knew for certain the group's purpose, or who or what they really were.[8]

A Georgia newspaper attempted to describe the Ku-Klux: "Their costume is said to be peculiar and significant . . . a white garment made up after the manner of a shroud, surmounted by a cap on the front of which is displayed a death's head and the cross-bones." A few days after this article appeared, the editor was visited by a mysterious messenger, who without a word, delivered a written note, which was published in the newspaper soon thereafter. The note stated that the Klan's objective was neither "to molest any one nor is it our intention to interfere with State or Federal authorities," and claimed that "any evil that may be committed by persons in disguise and under the name of the KuKlux would face the severest punishment."[9]

By the spring of 1868, the phenomenon had spread through the lower South, representing the full mobilization of a collective identity and was fully recognized by the press. "Secret societies are the natural offspring and inevitable result of Despotism, Military, Political or Social," one Tennessee newspaper declared, "When the recent acts created Military Despotism in the South, they also, by a law of nature, created secret societies in opposition." A Mississippi newspaper commented on the transformation, observing that the Ku-Klux "was the subject of joke and laughter and regarded only as a source of amusement; but in the twinkling of an eye, it has expanded into the huge proportions of a giant full of power and of passion."[10]

"The Southern whites have declared that they will no longer tolerate a policy which sinks them into servitude and makes a semi-barbarous race their masters," noted one South Carolina newspaper. "Our radical rulers forced their policy on the South, and already are we beginning to see what the consequences will be." The resistance was described as:

> . . . the logical result of two years of despotism lies now before us in the fact of a counteraction on the part of the whites. From one end of the South to the other a new secret society, known as the 'Ku Klux Klan,' has been organized in opposition to universal negro suffrage and negro rule.

The Ku-Klux movement in the South was explained as the natural result of the imposition of military rule and a new government based on black suffrage. Such a government would be "always resisted" and "never to be respected." The newspaper took a further step in expanding on the source of the Ku-Klux's mysterious power. The riders were "thought to be the spirits of the Confederate dead—who, at midnight arise from their graves, and roam about to avenge themselves upon the authors, black and white, of the insults and injuries heaped upon the land they died for."[11]

A pro-Republican New Orleans newspaper sounded the alarm in May, describing the Ku Klux Klan as a "lawless association," composed of "defiant and disfranchised rebels," whose purpose was to oppose the new constitution and government "to the knife, if necessary," and claimed that "if it increases as rapidly during the next month nearly every village and place of public resort in the State will have its representative Klan." A Pennsylvania newspaper published an excerpt from a Freedmen's Bureau official's report to the secretary of war, which concluded that "between the Loyal League, the Ku-Klux-Klan, the whites disfranchised and the blacks enfranchised, I think there will most probably be a very disorderly, and probably a very bloody contest." This report enjoyed a wide distribution throughout the South as well.[12]

As these articles indicate, a key element of this phase of asymmetric warfare is the information battlefield, in which Southern newspapers held the decisive advantage. By spreading stories of the Ku-Klux and building images of their mysterious origins and activities into the popular imagination, the resistance groups were able to magnify their strength and power, creating an image of a vast monolithic organization of inordinate power in which every non-Republican white man in the South was a member or supporter. Despite the efforts of pro-Republican newspapers to label the Klan as "vile traitors and assassins," among other things, the dominant image of an overwhelming and uncontained movement of resistance could not be overcome.[13]

The *Daily Phoenix* characterized the Ku-Klux in general as men who "want peace, but also justice." The Ku-Klux of South Carolina was not made up of "cutthroats or desperadoes," and while not accepted or approved by many people of the state, the paper admitted that nonetheless, "They are men of firmness and nerve who strike because they believe it necessary for the protection of life, liberty, and property."[14]

The long summer and fall of 1868, stretching into the spring of 1869, marked the zenith of the Klan as an effective instrument of asymmetric warfare. The Ku-Klux relied on security and unpredictability. The groups possessed local knowledge of people and activities, which allowed for the effective targeting of individuals. The Klan carefully passed on reports within the local community of what actions had been taken, who was threatened or attacked and why, and under what circumstances. This information was used to create the perception that those outside of the collective identity were completely isolated and vulnerable. Most of these tales of local Ku-Klux activities were spread by word of mouth, and wild rumors were also repeated that gained the power of truth.[15]

Although Klan violence and Klan-attributed violence would continue over the next two years, the asymmetric attack represented by the Ku-Klux had served its purpose, as indicated by the headlines in a Pennsylvania newspaper in September 1868, which proclaimed *"Ku-Klux Democracy,"* *"They Number Half a Million," "A Democratic Organization,"* and *"Prospect of a Civil War."*[16] An item published in a Vermont newspaper claimed that the Ku-Klux was estimated at 300,000 men, described as "oath-bound veterans ready to strike at any point necessary," who had complete control of the Democratic press of the South. "What does this mean," the article proclaimed, "if not a new rebellion?"[17]

This was the ultimate power of the weaker actor applying two asymmetric advantages. The first was surprise, creating the image of the Klan as a completely unorthodox, mysterious, and ethereal organization of collective resistance. The second was the ability to dominate the information battlefield, all intended to weaken Northern morale and will, while at the same time demonstrating a capability intended to further white Southern identity mobilization while marginalizing support for the state Republican governments.

The Republican governments faced with this completely unexpected phenomenon lacked the ability to unify popular support, especially among the loyal white majority and black majority counties of the state. The freedmen had to be protected at all costs to maintain the political viability of the new governments, yet their protection would require some form of coercive power, meaning vast expenditures on militia forces, the majority of which would have to be made up of politically reliable blacks. Loyal whites in most states (except Tennessee and Arkansas) did not control a particular geographical area, and most whites, of course, were the source of the resistance, so the only recourse was to arm blacks, which potentially would create far more dangerous problems to peace and order than the deployment of the militia would solve.

What Klan activity did throughout the South was expose the weakness of the Republican state governments and the inability of the Republican Congress to impose its will on the South and achieve its strategic goal of transformation through the establishment of the freedmen as a political partner and social equal. Nearly all of the Republican governments of the South made very little progress in suppressing the resistance.

Control of the local courts became a significant point of violent struggle. Whoever controlled the courts, controlled law and order in the county. Therefore, magistrates with the ostensible power to enforce and apply a broader concept of the law at the expense of white Southerners made them a significant target. Magistrates were the central figure at the local level who resolved disputes and punished lawbreakers. They conducted hearings and trials, tried minor offenses, reviewed cases, and administered justice according to their own peculiar understanding of the law and what local conditions would tolerate. Armed groups targeted Republican magistrates to prevent the courts from operating in order to delegitimize the state governments by highlighting their inability to enforce the law and maintain order. Intimidation, targeted violence, assassination, murder, and general mayhem created a sense of uncertainty and fear and placed the local and state government on the defensive.[18]

Asymmetric warfare relies on destabilizing violence intended to send a specific message to influence or intimidate.[19] The freedmen were the focus of the violence far more for what they represented, perhaps, than who they were. The freedmen were the political actors who held the key to power in the Republican state governments. Without their votes, the government had no chance of survival. By preventing the freedmen from voting and neutralizing the magistrate's ability to maintain law and order at the local level, the weaker actor could gain leverage over the stronger actor. During the election in 1868, the Klan reportedly shut down the Republican vote in eight counties of middle Tennessee. A freedman, Richard Moore, residing in Lincoln County, testified that he was attacked and ordered not to vote Republican. Despite the disguises and attempts to mask their voices, Moore knew for certain that the men were "near neighbors and all rebels."[20]

The Ku-Klux was certainly the most prominent group, but it was just one of innumerous groups acting in resistance to the state governments to negate the power of the black citizen as voter and overwhelm the local magistrates. There was a great deal of fluidity to the various violent activities that were reported through the years that have become part of the standard narrative of Reconstruction. Southern justice in and of itself was often brutally harsh, with whipping as a standard form of legal punishment for whites and blacks. Thus, the number of whippings reported throughout this period, while often quite troubling, were considered at the time to be the normal punishment for any number of offenses. At other times, criminals, mercenaries, and even idle youths were hired to frighten and intimidate or, in some cases, to commit murder. Significantly, organized military forces, either state militia or U.S. units, were never targeted. Any type of violent confrontation with these units could elevate into an unwanted open combat. A few reckless or foolhardy individuals sometimes threatened military encampments with gunfire and actions resembling preparations for an attack, but they disappeared rapidly at the first sign of any concerted reaction.

Whereas some groups acted very purposefully as vigilante enforcers to administer rough justice, others were engaged in nothing more than psychopathic violence. Some groups ranged outside their immediate localities, while others ranged close to home. Many were young men, reckless and unafraid, and often out of work. Some activities were conducted with military efficiency against a specific political target, others were just random rampages, and still others were attempts to intimidate and create disorder. Some groups pursued a form of what would be called today ethnic cleansing—forcing undesirables to leave an area or be killed. Within this realm of violence, organized Klan members conducted night raids, bandits and criminals preyed on the local populace, and secret societies and regulators challenged state and local authorities. At times, the military (mostly state militia, but also at times U.S. forces) skirmished with armed groups, but the soldiers were never sure who was who.[21]

In Louisiana, fraud and violence were commonplace during elections. The state had been under military occupation since 1862 and had been functioning under a constitution ratified by loyal whites in 1864. With the onset of Congressional Reconstruction, a new constitution had been ratified in April of 1868. State Republican leaders believed that educated free blacks and ambitious former slaves would provide the leadership to deliver the votes of the black majority that was intended to be the core of the new voting base, which, in partnership with loyal whites, would assure Republican control for decades to come. However, this formula failed as loyal whites were just as opposed to unlimited suffrage extended to freedmen as the disfranchised white majority.[22] Political violence emerged in 1868, most of it attributed to the Ku-Klux. In July of 1868, a parish official in Opelousas, Louisiana, wrote a letter to Gov. H. C. Warmoth appealing for a justice of the peace and sheriff to be appointed "to enable loyal men to obtain protection for life and property." He reported that "the K.K.K. are at work all over the parish, and nearly every night, with all kinds of nonsense to frighten and terrify the freedmen." He provided some additional information that he hoped would stir the political instincts of the governor. "If the [U.S. Army] commanding general had acted on my true and earnest representation of the condition of affairs in this parish," he wrote, "the republican party would have elected the whole ticket by at least 800 majority."[23] Another letter to the governor also contained an appeal for law and order through effective police or militia. Claiming that he was "guilty of no offense except that of being an active Republican," the writer expressed his sense of impending danger. "I tremble," he concluded, "when I think of the consequences of a revolution which may be started on short warning."[24]

In South Carolina, the Ku-Klux was active in 10 up-country counties, with the goal of disrupting the Union Leagues or Loyal Leagues and preventing freedmen from voting. According to a Klan oath printed in the Anderson, South Carolina, newspaper, the group's political intent was clearly laid out. The Klan existed to regulate the Republican Party, break it up, and strengthen the Democratic Party. They intended to kill or drive out

Republican leaders, and either prevent freedmen from voting Republican or convince them to vote Democratic. It boiled down to a simple formula: "force them, fight them, kill them, shoot them." Solomon G. W. Dill, a former Confederate soldier who joined the Republican Party and supported black suffrage as a member of the state constitutional convention, was killed in June. A black state representative and another black state senator were killed in Abbeville County in October.[25]

Several other highly publicized political murders were committed that were a reflection of the layers of violence that arose in the South during this time. George W. Ashburn was murdered in Columbus, Georgia, in March of 1868. Ashburn was the perfect example of the scalawag. He was a Georgian who had been such a strong opponent of secession that he organized and led a Unionist regiment during the war. After the war, he became a judge under military occupation, assisted activities of the Freedmen's Bureau, and supported Congressional Reconstruction and universal black suffrage. Elected to the Georgia state constitutional convention, he played a prominent role and, after the convention's work was completed on March 11, he returned to Columbus and attended a large rally of freedmen on March 30. After midnight, somewhere between 25 and 50 well-dressed men wearing masks broke into the boarding house where he was staying and shot him multiple times. Maj. Gen. George Meade declared martial law, and after determining that the city's authorities were not responsive enough to search for the murderers, he put an army captain in charge of the city, who then proceeded to arrest (and possibly torture) a number of suspects. No one was ever convicted.[26]

In October, James M. Hinds, a congressman from Arkansas was killed. Hinds had moved to Arkansas from Minnesota immediately after the war. A former attorney, Hinds participated in the state's constitutional convention in 1868. When Arkansas was readmitted to the Union, Hinds was seated as a Congressman. He declined to be renominated and instead sought to build popular support for the Republican Party. Traveling in southern Arkansas in October, he was met with open hostility. While Hinds was riding on his way to a speaking engagement, he was attacked and mortally wounded by the secretary of the local county Democratic committee, who had reportedly made death threats against Hinds earlier in the day. The attacker was never prosecuted.[27]

In the beginning of January 1869, the *New York Herald* noted that the Ku-Klux was causing such a state of terror "that the black and white republicans scarcely dare speak of them." In March, the judge of the Superior Court of Georgia issued a proclamation ordering members of the Ku-Klux to surrender to the sheriff of Robeson County or any other state law officer. Any state citizen was authorized to capture the outlaws and, in case of flight or resistance, were first to ask for surrender; if refused, the citizen was free to kill without facing any charges. The pro-Republican *Weekly Standard* in Raleigh, North Carolina, described the avowed purpose of the Ku-Klux was "to drive away all who are true to the Government," and summarized its effect: "its many murders and crimes have put the state into a fearful condition."[28]

A month earlier, another Raleigh, North Carolina, newspaper had declared that "a whirlwind of hate has swept over the South." The editor observed that "the cessation of the war did not bring peace to the people." He proceeded to explain the transition from war to warfare: "The passions of the battlefield were transferred to the political councils of the nation. Hate has striven against hate, and the sound of angry strife has filled the land."[29] This doleful assessment summarized the second phase of asymmetric warfare. The passions of the battlefield, embodied in the Confederate national identity, had now been combined with a Southern identity and mobilized for violent action. In essence, mobilized white Southerners became a nonstate actor, developing leadership and organization structures that employed rituals, symbols, memories, and cultural representations related to a Southern-Confederate collective identity. Reconstruction completely overthrew the central belief of Southern white men that true republican citizenship and government was entirely based on free and economically independent masters of households. The *Nashville Union and Dispatch* echoed this sentiment, asserting that "no great body of white people, as in this case, will long submit quietly to being deprived of the exercise of the highest attributes of citizenship." Completely contrary to the Republican ideology, white Southerners believed the protection of what was right took precedence over the protection of individual rights.[30]

Now either disfranchised or marginalized from the political process, these men had no faith that the Congressionally mandated state government would guarantee their security. The organization of Loyal Leagues and the purported clandestine nature of their activities, as well as their armed and unarmed public actions, served to confirm their greatest fears. Collective identity was far more important as a unifier and initiator of action than any notions of white supremacy. The Ku-Klux phenomenon emerged almost simultaneously with Union League political activity and the establishment of Republican-controlled governments.[31]

With a collective identity mobilized and threatened by a clearly defined threat of carpetbaggers and scalawags, backed by the power of newly enfranchised blacks, resistance moved into the second phase of asymmetric warfare. Violence was intended to accomplish several goals. First and most importantly, it was to deny the legitimacy of the Congressional Reconstruction strategy by delegitimizing the Republican state governments through preventing effective law and order at the local level. Second, it was employed to demonstrate that the weaker actor had power and the capability to sustain its efforts. In doing so, the resistance gained additional support, while encouraging others to cross the threshold from passive support to active resistance. Third, it was to neutralize the source of power of the Republican governments—the freedmen as the key political actor.

Layered violence was a vastly complex, fluid, and ambiguous condition that had substantial variations from area to area or state to state throughout the South. Layered violence, however, allowed the weaker actor to achieve multiple, overlapping, and sometimes mutually contradictory goals. It included the settling of local and private conflicts, achieving social control of freedmen, and imposing extralegal punishments. Although none of these had any relationship to the larger goals of the resistance, they contributed to the overall intended effect of violence as a process within asymmetric warfare.[32]

The political participation of the freedmen was the primary focus of the violence that served the goals of the weaker actor in its conduct of asymmetric warfare, making them legitimate targets of both lethal as well as non-lethal attacks. Overt political behavior or participation in specific activities, such as political organizing, joining the Union League, joining the militia, or organizing armed bands, removed from the participants any protection as innocents or neutrals. This ideological-based violence was intended to remove or neutralize freedmen as political actors and eliminate any potential armed threat.[33]

As an element of asymmetric warfare, the Ku-Klux appeared as a completely unexpected manifestation of resistance. The bizarre and outlandish costumes and performances hid a deadly serious purpose that took the state governments by complete surprise, and they seemed powerless to stop them. Their acts of violence were community sanctioned and effectively demonstrated that those who had been placed outside the collective Southern identity were completely alone, and neither the state nor the federal government had the capacity or power to protect them. The psychological effect was significant, as it raised the risks individuals were willing to take to continue to support the Republican governments to an unacceptable level. When it appeared that the Republican leaders could not make good their promises, the freedmen often retreated into passivity and neutrality. Ku-Klux activities contributed to a concomitant rise in general lawlessness and violence that was often attributed to the Klan. The cumulative effect of these layered levels of violence served its purpose of controlling the emergence of freedmen as Republican agents of change and reestablishing the dominance of white Southerners.[34]

In addition to this layered violence, the weaker actor almost immediately gained the decisive advantage on the information battlefield. Southern newspapers published and republished information that served to reinforce the sense of collective identity and resistance, while also battering the

Republican state governments with accusations of corruption and mismanagement. Republican idealism and rhetoric was no match for the power of
the terms and images employed by Southern newspapers. These terms,
most significantly "carpetbagger," were a powerful weapon of asymmetric
warfare that the weaker actor employed so effectively that the dominant
actor was never able to compete on this battlefield. After these terms took
hold in the popular imagination of both Northerners and Southerners by
1870, the Congressional strategy had been fatally damaged.[35]

By 1870, there were clear indications of a weakening in Northern resolve
to continue the effort to achieve the strategic goals of Congressional Reconstruction. The population of the North was bombarded with reports of the
disorder in the Southern states. The *Chicago Times* described the conditions as "organized hell." A. J. Fletcher, secretary of state for Tennessee,
described the Ku-Klux as a military organization that had the capability of
assembling anywhere from 80 to 100 men in a matter of hours to take
action. When it did take action, Fletcher said, "it acts with celerity and precision and rarely fails of accomplishing its object." He estimated upwards
of 40,000 men were associated with the Klan in his state alone and that
despite all of the violence directed against freedmen, Northerners, or loyalists, no one had been punished.[36]

The influential Radical Republican senator Oliver P. Morton of Indiana
highlighted that the Tennessee Ku-Klux operations were so effective that the
law was powerless. He described some of the reasons for their ability to maintain effectiveness: their use of disguises, the inability to gain any information
about their activities, the concealment of their operations, and the assistance they have among the populace. In addition, he described the Klan as:

> . . . a conspiracy in all the Southern States for the purpose of breaking down
> the Republican Party by deeds of blood and violence, by creating a reign of
> terror that shall induce Union men either to abandon the country or to
> abandon their principles and purchase their peace and safety by silence.[37]

Sen. Allen G. Thurman, a leading Democrat from Ohio, responded just
a few days later by attacking the assumptions of the Congressional Reconstruction strategy. "You put these states in the hands of loyal men," Thurman stated, "and in addition to this, you have had the military to help you.
Now, why is it that you have not had law and order and peace?" Thurman
highlighted the significant flaw in the Republican strategy—the assumption
that the new social and political order in the Southern states would result
in peace and prosperity. "You have tried," he said, "to bring about a normal
condition of things by military rulers, loyal legislatures, loyal governors, and
by all the machinery of loyalty." "Gentlemen," Thurman concluded in exasperation, "give it up." A Washington, D.C., newspaper displayed the same
frustration and focused on the source of the resistance that manifested

itself in the Ku-Klux phenomenon. "Five years of a peace policy have failed to convert the Ku Klux to a love of law and order. The asperities of the rebellion have not died out. They seem to be concentrated in the Ku Klux and to flame forth as blood-red as in the hottest period of war."[38] The asperities were related to the Confederate national identity and now combined with a Southern identity that had been reinvigorated by the Congressional strategy for Reconstruction.

The *Daily Phoenix* in Columbia, South Carolina, echoed these sentiments and summarized the attitude of many Southerners regarding the social and political conditions arising from the intent of Congressional Reconstruction to ensure Republican control of the Southern state governments that would guarantee neither peace nor order. "The radicals have not freed our slaves," the article asserted. "They have not given the negro his liberty, and his power [is not] for his benefit, but for their own. He is much their slave now as he was ours in the olden time—the form only of his bondage has changed."[39]

The Marianna, Florida, *Courier* gave a much clearer voice to the spirit of resistance contained in the collective identity of white Southerners, illustrating a facet of self-definition as the source of mobilization:

> We can never regard the whole of this reconstruction infamy in any other light than the lowest, most debased and dastardly piece of cruel and unmitigation [sic] tyranny the vicious brains of cowardly oppressors could conceive to wreak vengeance upon a downfallen though proud and noble people.[40]

Taken aback by the intensity and unorthodox nature of the resistance, the Republican Congress sought to settle the situation through another amendment to the Constitution. In Congress, debates over the Fifteenth Amendment indicated that the Republicans sought to secure the right to vote for black men throughout the nation, sincerely believing that they deserved the ballot through their inherent love of the Union, and, in turn, the ballot would serve as the tool to create responsible citizens who would protect it from abuse. The Republican Congress, holding the majorities in both houses, sought to craft a properly worded amendment that would be accepted by the majority of the states. The amendment as it was submitted prohibited denying the right of suffrage to anyone on the basis of race, color, or previous condition of servitude. With this amendment, Republicans were confident that they had essentially achieved the purpose of the war and that the amendment would bring about reconciliation and peace, while arming blacks in the South with the means to defend their rights. Thus, by placing their faith entirely in the power of the Constitution, the Republicans in Congress mistakenly believed they were achieving the goals of their strategy.[41]

A Memphis, Tennessee, newspaper commented that the Radicals seemed bent on forcing universal suffrage on the South for the freedmen but preventing the same right for former Confederates. "They are trying to govern the country after their ideas, regardless of the wishes or interest of the people down here, as if we were mere children." A South Carolina newspaper declared that the Republican Party was "regarding Congress as the sovereign power" within the United States.[42]

Although the violence of the second phase of asymmetric warfare would continue, it had already served its purpose. It was now a backdrop to a concerted political opposition effort by state Democrats to cripple fatally the already politically reeling Republican governments. Georgia provides an example of how one state government responded to the shock and surprise of the violent resistance now mobilized against it. The Georgia legislature formed under the Reconstruction Act met in June 1868 and ratified the Fourteenth Amendment as required, resulting in Congress readmitting Georgia to the Union. The new governor was Rufus Bullock, a native of New York who moved to Georgia for business reasons in 1860. He had been a commissioned officer in the Confederate army and supervised telegraph and rail services for the Quartermaster Department. In the state constitutional convention ordered by Congress, he played a prominent role, and the Republicans nominated him for governor.[43]

But Bullock faced significant resistance from the state's Confederate political elite. Robert Toombs, the senator who led Georgia out of the Union in 1861 and who had served briefly as the secretary of state for the Confederate government and then as a brigade commander in Robert E. Lee's army, was allied with Howell Cobb, a former governor before the war and a major general in the Confederate army. Both men were completely hostile to Congressional Reconstruction, and their voices were very influential in rallying opposition to the new government. As previous leaders of the state both prior to and during the war, they had a loyal following of former Confederates. This support extended to poor whites, whose loyalty the Republicans had hoped to capture. One important assumption in the Congressional strategy was the belief that common Southerners perceived themselves as victims of fire-eaters like Toombs, who brought on secession and war. In the aftermath of defeat, Republicans were confident that these plain folk would join with the loyal whites in supporting a new state government that had excluded these leaders from power. This was a seriously flawed assumption, as it ignored the role of collective identity, and, as a result, this did not happen to any extent in the South. The Republican governments remained dangerously separated from the white majority, who continued to take their promptings from the representatives of the Confederate wartime leadership.

Bullock's attempt to meet the Congressional goals by building a biracial coalition dedicated to economic integration with the North quickly ran into trouble. The governor's extensive contacts with Northern businessmen and investors led him to promote a number of costly state initiatives intended to move Georgia rapidly into the North's sphere. Bullock's willingness to cut corners led his opponents to charge him with corruption and malfeasance. In September, Bullock's fragile political alliance fell apart. A new alliance of moderate Republicans and Democrats challenged the entire legitimacy of the state government, arguing that the Fourteenth Amendment had given black men citizenship, but this new status did not confirm on them the right to hold elective office. The legislature therefore declared the 28 newly elected black legislators ineligible until the state supreme court made a decision on their status and replaced them with candidates—all white men—who received the second most votes in the election.[44]

Shortly after the legislature's action, one of the deposed black legislators joined with two white Union army veterans—one had arrived in Georgia in 1865 and established himself as a planter, and the other was a Republican Congressional candidate and a Freedmen's Bureau agent—to lead a march to the Mitchell County courthouse at Camilla to join a Republican political rally. Here was one of the most visible examples of everything white Southerners feared. Two white carpetbaggers, Union army veterans who had despoiled Georgia during Sherman's infamous "march to the sea," had been transformed: one was seeking political office, and the other was an agent of the government. They were joined by a black politician, voted into office because the franchise had been denied to the white majority. Accompanied by a bandwagon, they led a large group of blacks, a number of whom were either armed or were carrying sticks. It appeared to be a nightmare come true.

The county sheriff met the crowd outside of town and refused to allow armed men to continue. They proceeded nonetheless, and as they entered the town, they were confronted by an armed posse of white men intending to keep order. Tensions were very high, and when a drunken white man fired shots at the bandwagon, a general gun battle broke out. The crowd ran for safety into the woods, and the posse pursued the escapees for several days, shooting down every individual they encountered. In the aftermath, 9 freedmen had been killed, and at least 25 had been wounded. Although thoroughly investigated by U.S. military officers and Freedmen's Bureau agents, no one was ever arrested or tried for this quasi-legal killing spree.[45]

This event indicated the extent of Southern white fears that blacks in politically organized groups possessing arms and led by white Republican Radicals would lead to violence and escalate into the wholesale slaughter of white Southerners. This incident was highly publicized, and for the Republican governments, it became a clear warning sign of what might happen if Union League activity or any other political organization efforts of blacks became too visible.

In March 1869, the anti-Bullock alliance in the legislature rejected the Fifteenth Amendment. To add insult to injury, Georgia's electoral votes went to the Democrats in the 1868 presidential election. This was an intolerable situation, but Congress had no answer until Governor Bullock made a visit to Washington and convinced President Grant that what policy Congress could make regarding Reconstruction, it could also unmake. Bullock proposed that Congress declare its previous readmission of Georgia as invalid and return the state to military control. Congress, ironically imitating the Georgia legislature, declared the representatives of the state ineligible, and the commander of the Department of the South (Georgia, Florida, Alabama, North Carolina, and South Carolina), Maj. Gen. Alfred Terry, was declared the commander of the district of Georgia. Given the authority of district commanders under the Reconstruction Acts, General Terry then returned martial law to the state in December 1869. Terry had submitted a report on the security situation in Georgia in August to the secretary of war, claiming that violence had left large parts of the state ungoverned. Law officers were either controlled by intimidation or in league with the Ku-Klux and other "insurrectionary organizations." Now, with total power but with little guidance from Washington and very little assistance from the attorney general, Terry felt his way along, using a board of officers to determine eligibility of candidates from the previous election and dispatching troops to support civil authorities and oversee the polls, but the violence had subsided.[46]

Terry took the least controversial approach, in essence turning back the clock to 1867 and restoring the state legislature to its original composition (including the previously elected black members, who had been validated to have a constitutional right to hold office by the Georgia Supreme Court) and ensuring that there were no factions capable of interfering with Congressional mandates. This reorganization would align the necessary number of Southern states needed to ratify the Fifteenth Amendment. North Carolina, Louisiana, South Carolina, Arkansas, Florida, Alabama, and Mississippi, all reliable Republican governments, had already ratified. Texas ratified in March as a precondition for readmission. President Grant, in his proclamation addressed to Congress, called the ratification of the Fifteenth Amendment as "the most important event that has occurred since the nation came to life."[47] Others were not so enthusiastic. The amendment was called "a deliberate and indubitable fraud," "unconstitutional, unjust and oppressive," and "palpable and flagrant . . . coercion."[48]

By July, Congress had readmitted Georgia once again into the Union, and its representatives were seated; Congress also allowed new elections for the Georgia General Assembly to be held under the provisions of the accepted state constitution. The Democrats won commanding majorities in both houses in the December 1870 elections for the next general assembly that would convene in November 1871. Bullock attempted once more to return the state to Terry's control, but after Congress recognized the validity of the election results, he fled the state in late October 1871 to escape impeachment. In a special election held in December, Democrat and ex-Confederate colonel James M. Smith was elected to complete Bullock's term. Consequently, as of January 1872, Georgia was fully under the control of the Redeemers, as the state's resurgent white conservative Democrats came to be known.

Georgia had returned to the Union, but it was no longer under Republican control. The asymmetric approach of violence by the mobilized collective identity created a condition that Governor Bullock attempted to forestall by an unorthodox maneuver of his own—the restoration of martial law. Although this legally questionable action worked, he was unable to preserve his existing coalition and failed to delay the general assembly elections. As a result, the Democrats carried the state and assumed power.

In Tennessee and Arkansas, Klan activity served as the backdrop to political realignments that doomed the Republican governments. For these governors, the attempt to use force to quell the violence was fraught with political danger. The dilemma the Republican state governments faced was the decision to impose control through force—usually requesting the employment of federal troops, or more dangerously, by ordering the state militia to take action. The largely symbolic employment of the militia represented an effort of the Republican governments to assert legitimacy and establish security. The action often backfired, as the militia was immediately labeled as a tool of despotism, intended to overawe opposition. Black militia units were especially resented and feared, creating an even greater level of mobilization and resistance. The establishment of the militia for providing security for Republican supporters only increased fears among white Southerners, who then formed armed groups for protection, creating a series of fear-driven actions and counteractions that spiraled into violence.[49] The real danger, however, was the political backlash that inevitably followed the militia's deployment. The governors of Tennessee and Arkansas were the most vigorous in using militia forces, but the inability of these forces to accomplish anything substantial belied their utility and led to a serious political crisis in each of these states.

The Republican government of Tennessee, under Gov. William G. Brownlow, whose state had already been admitted to the Union prior to the Reconstruction Act, had many advantages for meeting the goals of Congress for Reconstruction. It had a large population of loyal whites in east Tennessee, and the state constitution had disfranchised former Confederates and excluded them from power. The one drawback was that as a fully reconstructed state readmitted to the Union, the governor had no federal military forces at his disposal. General Grant had made his policy clear that the army's role was to prevent conflict and preserve peace, but any employment of force would be used only to put down hostile mobs. Thus, Governor

Brownlow had to rely on the State Guard, made up primarily of white volunteers from east Tennessee, to enforce peace and order.

In 1867, the state militia force convinced the legislators that it had suppressed the Klan, even though it had spent most of its time patrolling uneventfully through the countryside. The State Guard was disbanded that year. By July of 1868, Governor Brownlow was facing a rising tide of resistance in the central and eastern sections of the state, where commissioners of voter registration, Union League members, and organizers were particularly targeted, declared that the goals of the Ku-Klux were to overthrow the existing government, abolish the suffrage of freedmen, and restore the franchise to "every rebel who fought to destroy the government." Brownlow sought to have Klan members declared outlaws and, when captured, face immediate execution.[50] An anti-Klan bill was passed in September that defined a Klansman as someone who associated with "any secret organization" who went about "by day or night disguised or otherwise, for the purpose of disturbing the peace." While no one could be shot on sight as Brownlow desired, individuals convicted faced a fine of $500 and five years' imprisonment, plus the loss of voting rights and disqualification from holding any elected office.[51]

In 1869, Governor Brownlow, also acting as the commander in chief of the state militia, declared martial law in nine counties (five counties known to be Ku-Klux strongholds were avoided). Civil courts were closed, and the militia made arrests from lists of names provided by the state. The mission of the militia units was ostensibly to uphold the law and bring peace, but, in reality, they were to make reconstruction work, essentially serving as Brownlow's means of political self-preservation.[52] Units patrolled the counties but encountered no significant resistance. Frank McCord, the editor of the *Pulaski Citizen*, left the state. Brownlow's personal political ambitions were stronger than his interest in preserving order as the militia commander. He eagerly accepted the state legislature's appointment as senator and left for Washington, leaving his lieutenant governor, DeWitt C. Senter, to continue the mission. Instead of pursuing Brownlow's policies, Senter reopened the civil courts, ordered the militia to cooperate with law enforcement, and ended martial law. Shortly afterward, he insisted on the repeal of the anti-Klan law, purged a number of Radical election officials, appointed moderates who would allow former Confederates to register to vote, and demobilized the State Guard. Senter was reelected in August of 1869 by a coalition of Democrats and conservative and moderate Republicans. Bolstered by white Southern votes, the Democrats took control of the state government in November 1870 and began to dismantle the state constitution that represented the goals of the Congressional Republican strategy. John Calvin Brown, a former Confederate general and Ku-Klux member, became the new governor. The 1870 constitutional convention extended the franchise to all citizens at least 21 years of age who paid a poll tax, and it limited the governor's ability to call the militia.[53]

The rapid fall of the Brownlow Radical Republican model indicated how fragile the Republican governments actually were and how powerful the mobilized collective identity of white Southerners was in providing a level of resistance that could not be overcome. The asymmetric challenge of the mysterious Ku-Klux and its image of all-pervasive power and potential military capability led Governor Senter to seek accommodation and accept a new power relationship rather than risk a possible open war or a sustained level of violence that would lead to perpetual disorder. Just as Tennessee's Reconstruction government had faced widespread resistance, Gov. Powell Clayton of Arkansas also was confronted with a similar asymmetric warfare challenge.

Powell Clayton, a native of Pennsylvania, joined the Union army in 1861 as a captain. By the end of the war, he was a general commanding a cavalry division. Clayton's unit spent most of the war in Arkansas, where it had experience in combat against Confederate forces as well as irregulars. In 1867, Clayton became a prominent leader in the newly formed Republican Party, first as a delegate to the state convention, and then as the party's nominee for governor. Clayton was inaugurated governor of Arkansas in July 1868. Under the new Republican constitution, the governor had the power to appoint a number of county officers, and Clayton immediately set out to solidify the state's control and authority by appointing loyal men to these positions.

With the onset of resistance shortly after taking office, Clayton, like Governor Brownlow of Tennessee, nullified votes cast in certain areas contested by the Klan and used the state militia, composed of loyal whites and freedmen, to suppress the growing threat of the Ku-Klux. He declared martial law in 14 counties, mostly in southern and eastern Arkansas, and employed the militia.[54] Clayton opted for the state militia over regular U.S. troops. As one Memphis newspaper noted, "the United States troops do not know the people and the country, and the people behave while the United States troops are in the vicinity, and break out worse than ever when they leave."[55]

Clayton's force, however, was a militia in name only. Lacking uniforms, equipment, and supplies, the men spent most of their time confiscating food. Unlike other instances in other states where all signs of resistance disappeared with the arrival of the militia, in Arkansas, the militia was met with armed resistance from county residents. There were skirmishes and several casualties. Reinforced by a black militia unit, the state force moved through the counties under martial law. Several individuals were arrested under suspicion of belonging to the Ku-Klux, and one man was executed on the spot for murder. Reports of looting, rape, and wanton destruction began to fill the newspapers of the South as stories of undisciplined black militia units roaming the countryside became a commonplace image that undoubt-edly hardened many attitudes.[56] A commentary in the *New Orleans Crescent* spoke for many Southerners: "These outrages are all in the name of 'loyalty'— all designed to strengthen the government and illustrate the beauties of reconstruction as ordered by Congress and executed by the carpet-baggers and scalawags."[57]

In one Arkansas County, a home guard was formed to establish order to avoid the governor's order to deploy the black militia. The arrest, conviction, and execution of a Klan member seemed to indicate some level of effective-ness that kept the black militia units from occupying the county. In another county, a militia unit took hostages and threatened to kill them and destroy the town of Augusta unless the Klan ceased activities.

When black militia units jailed several suspected Klan members, a num-ber of skirmishes took place as armed groups attacked the militia in an attempt to free the prisoners. The battles escalated as both sides brought in reinforcements, but the militia successfully warded off the attacks. In the aftermath, one Klan suspect was convicted of murder and executed; the others were shot while trying to escape, a standard explanation for any extra-legal killing in the South. The depredations of the militia were severe enough that one county under martial law asked for federal troops to con-duct operations. By 1869, martial law had been lifted, and Clayton had gained at least a limited breathing space to establish a stable government.[58]

Clayton's approach of direct military action through raids and sweeps resulted in open firefights, indicating that Klan-attributed violence was more general lawlessness and anarchy rather than an attempt to neutralize the Republican government. Certainly, the willingness of some paramilitary bands to resist the militia may have been related to antigovernment action, but the fact that county self-defense forces could be raised to bring order was a clear indication that Klan activity could be controlled. Governor Clay-ton, like Tennessee Governor Brownlow, in their aggressive use of militia forces, created an image of government rule by bayonet in the minds of South-erners; moreover, any state governor after 1869 would find it very difficult to call up a militia, especially a force largely composed of blacks.[59]

In Arkansas, Clayton's hold on power began to unravel as a splinter group of Republicans sought to end statewide corruption and return the franchise to former Confederates. Led by the lieutenant governor, James Johnson, this faction sufficiently weakened the party in the 1870 elections that the Demo-crats, despite disfranchisement, made significant gains in the state legisla-ture. The stage was set for a political reversal of fortune, driven indirectly by violence and disorder that hampered Clayton's ability to establish a sta-ble government.

In April of 1871, the former governor and now Senator Clayton gave a self-serving speech before an appreciative crowd in Washington, D.C., presenting his own version of events as governor. He claimed that he faced 30,000 Ku-Klux members in an organization so powerful that much of the state functions were paralyzed or seriously impaired. As a "last desperate resort," Clayton claimed, he declared martial law and organized a force of 2,000 men that resulted in the arrest and trial by military commission of "several members" of the Klan.[60] The effort of 2,000 troops that resulted in the capture of just a few individuals and the execution of several others was nothing to brag about; it does, however, point to the power of the weaker actor to use its asymmetric advantage to shift and reorient as circumstances dictated, damaging the cohesiveness of the governing coalition that sustained the state governments.

In 1872, the Democrats supported Joseph Brooks, the liberal Republican candidate for governor, despite his strong support of black political rights and the fact that he had been a target of the assassination attempt in 1868 that killed Congressman James Hinds. The other Republican candidate was Elisha Baxter, a Unionist refugee during the war who returned to Arkansas with the advance of the Union army and was appointed a judge in the Clayton administration. The 1872 election was irregular enough that Congress invalidated the state's electoral votes. Despite the irregularities, Baxter claimed victory. Baxter offered to end the disfranchisement of former Confederates as a means of consolidating support, and, in 1873, Baxter's victory was accepted by a wide majority. This led to a Democratic-controlled state legislature, which immediately began to call for a convention to revise the state constitution. Thoroughly alarmed, the Clayton Republicans joined ranks and obtained a ruling that declared Brooks the legitimate governor.[61]

Brooks gathered an armed entourage and forced Baxter out of the governor's office. Baxter gathered his own armed supporters, and President Grant found himself in the unusual position of using U.S. troops to prevent violence between Republicans. The state capital of Little Rock became an armed camp, and factions battled each other in the countryside. Two companies of U.S. infantry were in Little Rock, monitoring the streets to prevent violence, while maintaining complete neutrality. The crisis ended with a finding by the U.S. attorney general that led to a presidential proclamation recognizing Baxter as the legitimate governor in 1874.[62] The new state constitution (still in effect today) severely limited the power of the governor and dealt with corruption and debt, the greatest legacies of the Republican-controlled state government.

Like Brownlow in Tennessee, the attraction of a Senate seat pulled Clayton, the central figure of the Congressional Reconstruction plan, away from the state. Although the violence in Tennessee and Arkansas followed completely different patterns, the similar decisions to employ the militia, which was intended to suppress the resistance, both failed to achieve results. The offer of Governor Senter in Tennessee and Governor Baxter in Arkansas to end disfranchisement of former Confederates changed the situation entirely, making the asymmetric approach of violence unnecessary and allowing the Southern mobilized collective identity to transition to political action.

In Florida, the coalition Republican government was far less willing to implement the Congressional goals of social and political transformation, which had a fundamental effect on the direction of this state during Reconstruction. The moderate Republicans, who held the balance of power, had more interests in common with the marginalized white majority. They were more interested in advancing mutual commercial interests than in revolution and punishment of traitors. This accommodationist tack largely moderated the government's approach to black suffrage and equality. Congress accepted the state's far less ambitious constitution and admitted Florida in 1868. Harrison Reed, the new governor, appointed former Confederates and Whigs to both his cabinet and to the circuit court. Reed had little or no interest in remaking the state, asserting that nominally allying with the Republicans in Congress gave Florida access to federal largesse. Reed's coalition collapsed almost immediately, as blacks were dissatisfied with unfulfilled promises, a Democratic minority was serving as a constant irritant, and moderate whites were moving toward ousting the governor by impeachment.[63]

In 1867, Florida Freedmen's Bureau agents were organizing clandestine organizations called the Lincoln Brotherhood. Like the Union Leagues, they were intended to organize freedmen for political action. Lodges were formed in most counties, and members swore an oath to protect and defend the Constitution, rescue lodge brothers in trouble, and vote for Republican candidates.[64] By April 1868, Klan notices began appearing, and Young Men's Democratic Clubs and armed groups called regulators were active in the northern and central part of the state. In November of 1868, a cache of arms and ammunition destined for the state militia was intercepted and destroyed by unknown raiders. Federal troops were dispatched across northern Florida. Two companies occupied Jacksonville, and a number of detachments occupied the major towns. Jackson County served as the main arena for violence in Florida during this time. The counties that experienced the most violence had black majorities along the Georgia-Florida border; in contrast, counties with few blacks had little violence. Jackson County provides an example of layered violence that was largely asocial and highly symbolic, born from the destruction and hatred as a result of the Union incursion into the county in 1864 and fostered by a lack of adequate law enforcement. A generation of men too young to fight then, but hardened by the wartime experience, were old enough now to take matters into their own hands. Violence in this county was localized and often intensely personal, resulting from animosities and divisions between pro-Unionist and pro-Confederate factions within the county.[65]

At the onset of Congressional Reconstruction and the enfranchisement of blacks, small groups of armed young men long prone to violence and often accompanied by hired guns, rode about causing fear and mayhem, attacking individuals because of what they represented rather than who they were. Individual reputation and character meant nothing if someone was declared a carpetbagger or a Radical. Those individuals were instantly the enemy. Even the arrival of a federal detachment in the county seat of Marianna had little effect. County Republican leaders and Freedmen's Bureau agents active in organizing blacks were threatened, challenged openly, and physically attacked. Most were killed in ambush, while those suspected of initiating those attacks were themselves the targets of retaliatory ambushes. One criminal gang responsible for much of the violence finally went too far. A number of freedmen became targets of especially shocking murders that led to indictments and the flight of the perpetrators to avoid justice.[66] In other Florida counties, blacks suspected of having influence were targeted, and sheriffs were harassed and threatened; old scores between Unionists and Confederates from the war were settled, and Republican legislators were always in danger. Facing the real prospect of death, some legislators resigned when threatened, which opened opportunities to replace them with Democrats. This kind of violence ebbed and flowed throughout the state between 1868 and 1871.

A *Richmond Dispatch* reporter in Florida in 1869 tarred Governor Reed as a carpetbagger who "has only one merit—that of timidity." He proposed "to call out the negro militia in imitation of the Governors of Arkansas and Tennessee," but stopped short when it became clear that such a step would lead to armed resistance.[67] The stories of militia rampages and destruction from Arkansas remained fresh over the next three years and served to galvanize opposition and resistance, as well as to deter governors from taking this option.

Rather than use the militia as other governors had done, Reed negotiated a settlement in Jackson County, getting approval for his appointments of county officials from the white leadership. Although not a comprehensive solution, it did bring about a substantial decline in violence in the county. Elsewhere, as Democrats increasingly took control of the county governments, the violence also declined.[68]

The 1870 state elections were a particularly overt demonstration of force in establishing a balance of power between the factions at the state level, while Democrats established or maintained a level of local control. There were few acts of violence, but armed men stationed at polling stations and riders galloping through the countryside were sufficient to keep freedmen from the polls. Confrontations were commonplace, and weapons were freely brandished. J. J. Dickison, a former Confederate cavalry officer known for leading guerrilla raids against Union forces in Florida during the war, led armed cavalry troops through groups of voters. On the eve of the election, Governor Reed appealed to President Grant for five companies to preserve peace at the polls. Reed would later claim that he faced revolutionaries who were using assassination and murder in preparation for an imminent armed assault on the capitol. Grant was reluctant to respond but eventually acceded to the request far too late for the War Department to respond in time.[69]

Shortly after the election, two companies were dispatched. At the local level, especially in Jackson and Columbia Counties, the pattern of killings followed by revenge killings over the years led to the periodic deployment of U.S. troops, but these efforts were largely ineffective; the killings had already taken place, or targeted individuals had already been forced to leave the county. Ironically, the Enforcement Act, passed by Congress to deal with widespread violence in South Carolina, was actually more effectively used in Florida than anywhere else in the South. Cooperating with state officials, federal law enforcement officials were able to obtain indictments and prosecutions for certain lawless individuals without resorting to martial law or suspending the writ of habeas corpus.[70]

As Republican power waned amid factionalism, Reed faced and survived a number of impeachment votes, finishing his term in 1873.[71] Ossian Hart, a Unionist slave owner who had avoided Confederate service with a medical exemption, was elected governor despite a challenge from a fusionist liberal Republican-Democrat challenge. Hart attempted to change the political balance of power by announcing the removal of certain county officials and replacing them with loyalists. In Lake City, in Columbia County, the governor's attempt led to a dramatic display of resistance. Under cover of darkness, a group of armed men systematically killed the new appointees, moving from house to house and firing into the homes. Hart responded forcefully, deploying black militia and risking an open battle. Rather than pursue an entirely coercive approach, Hart combined it with conciliation and compromise. The governor and his entire cabinet went to Lake City to urge peace and order. He also worked with prominent local Democrats to gain their cooperation in prosecuting the assailants, while stepping away from making new appointments.[72]

This incident indicates the uniqueness of the situation in Florida. It was essentially an armed peace, with Republicans maintaining state power through the votes of freedmen; as long the moderates remained in control, Democrats were willing to accept the status quo and enjoy the benefits of the economic transformation. But any overt action to challenge the balance or any perceived attempt to gain a political advantage was met with a dramatic act of murder and assassination. In 1875, for example, state senator Elisha G. Johnson was killed in an ambush. Johnson, a prominent Republican leader in Columbia County, had been accused of election fraud. His death gave the Democrats a majority in the state senate.

Florida also benefited from being the vanguard of what would become known as the New South, as tourists, investors, immigrants, developers, and individuals seeking to resettle arrived in increasing numbers to take advantage of the mild climate, diversified agriculture, cheap land, and timber resources. White Floridians of all political factions welcomed the influx of these newcomers, and these new conditions helped to mitigate disharmony and create an alliance of convenience that moved the state toward stability and a departure from the Congressional Reconstruction goals.[73]

Hart died in office in 1875, and Marcellus Stearns became governor. Stearns was a former Union soldier from Maine who remained in Florida at the end of the war and served as a Freedmen's Bureau agent. Stearns had been a delegate to the 1868 constitutional convention and was a member of the new state assembly until 1872, when he was elected lieutenant governor. While his background would mark him as a Radical in the mold of the desired leader the Congressional strategy required, he was actually quite reticent to follow that path. In fact, Stearns had played an important role to ensure that the state constitution limited black representation to assure white control. Stearns was only 34 years old and worked with the Democrats who now controlled the state assembly, balancing the budget and fostering the emerging New South spirit.

The role of violence (or the threat of violence) is an essential aspect of asymmetric warfare. The weaker actor must employ violence in such a way as to gain the initiative through surprise. The Ku-Klux achieved this in measures far greater than anyone imagined. The phenomenon swept the South and was so effective that it continues as part of the American collective consciousness. The unorthodox approach and its image of power and success was the asymmetric advantage that allowed a mobilized collective identity to cross the threshold of support and sustain resistance. Ku-Klux violence spawned many imitators who took advantage of the weak control of the Republican state governments to spread mayhem. This mixture of purposeful, social, and asocial violence represented layered violence that served the purpose of the asymmetric actor to demonstrate the inability of the state governments to defend the key political actor that guaranteed them power—the freedman and his ballot. The secondary target was composed of the white allies of the freedmen, local magistrates, and law enforcement officials who had the power to control local affairs in the state. By neutralizing these targets, the state was made powerless to take any other action except use of armed force. But this option, as demonstrated in the futility of the Tennessee State Guard and the militia forces of Arkansas to restore the dominant actor's control to any degree, resulted in a political compromise, giving disfranchised white Southerners the vote and thereby ensuring the collapse of the Republican state governments.

Unlike other states, the accommodation between moderate Republicans and marginalized whites in Florida provided the means for relative stability.

The violence was mostly localized and, while directed at perceived enemies, was typically asocial and served no larger purpose. The intimidation and threats that marked the election of 1870 in an effort to limit the votes of freedmen were largely unsuccessful, driving Democrats toward solidifying an alliance with Republican moderates and liberals. Although there were a number of highly publicized acts of murder and assassination, they were rare representations of the latent power of Democrats to maintain the balance of power through violent action.

As these events indicate, the Republican strategy, battered on the information battlefield and losing the initiative in the face of an unexpected and unorthodox resistance, was already beginning to falter by 1870. To save it would require some significant effort from Congress and a demonstration of federal power and authority.

6

Asymmetric Warfare Phase II: The Dominant Actor Responds (1870–1873)

Confronted with a completely unexpected form of resistance and violence in the form of the Ku-Klux, Congress was caught by surprise, but the leading Republicans had difficulty in mustering the collective political will and interest to formulate an appropriate response. Between 1871 and 1872, the imaginative power of the Ku-Klux was at its height. "We have reports of such Southern scenes of violence and bloodshed almost every day," the *New York Herald* declared, despairing that further coercive measures by Congress would assist the demoralized Republicans in controlling the Southern states. "After such a four years' war as that of our late 'Southern rebellion' can it be expected that the people subdued can quietly adapt themselves in five years, ten years or twenty years to a revolution which has torn up their political and social system by the roots?" A Washington, D.C., newspaper echoed these sentiments, after reviewing the current state of affairs in the South. Every part of Tennessee, middle and southern Kentucky, Louisiana, Texas, northern Mississippi and Alabama, and western North and South Carolina were strongholds of the Ku-Klux. Smaller pockets existed in Arkansas, middle and southern Mississippi and Alabama, Georgia, Florida, and southern Virginia. Two-thirds to three-quarters of all white men ages 16–50 were members. "Southern leaders have never ceased to contemplate the possibility of their reinstatement in political power, and their ability to undo nearly all that the war accomplished." This resilience and growing confidence was reflected in the power, influence, and resourcefulness of an invisible enemy that had seized hold on the public mind, shaping perceptions and attitudes, and draining support away from the Congressional Reconstruction strategy. In Mississippi, for example, the resistance attributed to the Klan was so strong in the eastern counties of the state that, by 1870, the Republican government sought to find some workable political compromise to mollify former Confederates, while still protecting the political and civil rights of the freedmen.[1]

As the resistance mounted, Congress attempted to retake the initiative. Sen. Charles Sumner led the effort to maintain Congressional control over its Reconstruction strategy by placing additional restrictions on the remaining Southern states still not returned to the Union. The Virginia and North Carolina constitutions were accepted, but with the requirement that the state constitutions could not adopt amendments that limited basic rights related to voting and holding office. Although the Radical Republicans still directed the strategy of Reconstruction, there was growing concern that implementing the Congressional strategy was going to be more difficult than anticipated.

Congress assembled a Senate Select Committee to investigate the rising disorder in the South. The committee issued a report in March, 1871, titled *A Report on the Alleged Outrages in the Southern States.* It took testimony from a number of witnesses over several months, most of them from North Carolina, who created the image that no loyal man could openly express his political opinions for fear of retribution; loyal whites and freedmen were completely unprotected from attack; and fearing the midnight assault and threat of assassination, the same were under tremendous pressure to abandon their support for the state government.

The report, as summarized by a Pennsylvania newspaper, described the Ku-Klux as an "oath-bound secret organization" working only at night, whose sworn purpose was to "put down" the Republican Party and put up the Democratic Party in a "premeditated and determined scheme" to carry the entire South's electoral votes for the Democrats in 1872. Ohio congressman Job Stevenson, in a speech in Delaware, Ohio, a few months after the report was published, told his audience that 30,000–50,000 citizens in the South had been "outraged" since 1867. The report made it clear, as Stevenson claimed, that the "the Ku-Klux is the most powerful political machine ever invented. It has conquered several States, and may, if not checked, rule the whole country."[2]

Given this kind of impression, the report, issued in March, became the impetus for legislation. An Enforcement Act was passed to take action to protect black voters from intimidation, discrimination, or any attempt to prevent them from exercising the right to vote. The other action was the

Ku Klux Klan Act of April 1871. Also known as the force bill, it made certain individual activities federal offenses, such as denying a U.S. citizen the right to participate in political activities, own property, vote, or serve on a jury. It was also illegal to conspire or go in disguise with the purpose of depriving anyone equal privileges or immunities of the law. Congress put substantial teeth into the law, authorizing the president to intervene in state affairs and, at his discretion, selectively suspend the writ of habeas corpus. It also allowed U.S. forces to assist civil authorities in restoring law and order. Wendell Phillips, the abolitionist and vocal supporter of the Radical Republican strategy, recommended that to show its resolve, the government should hang a selected number of former Confederate generals to quell the Ku-Klux activities and ensure a Republican victory in the 1872 presidential election. Job Stevenson advocated the reimposition of martial law on the Southern states. "Better rule by the bayonet," he said in a speech, "than have them [the Ku-Klux] rule by the revolver."[3]

On May 15, the War Department distributed the proclamation by President Grant regarding the enforcement of the provisions of the Fourteenth Amendment. It referred to "persistent violations of the rights of citizens of the United States by combinations of lawless and disaffected persons in certain localities lately the theater of insurrection and military conflict." It called for "people of those parts of the country to suppress all such combinations by their own voluntary efforts through the agency of local laws and to maintain the rights of all citizens of the United States and to secure to all such citizens the equal protection of the laws." Grant added an almost plaintive plea for law and order:

> It is my earnest wish that peace and cheerful obedience to law may prevail throughout the land and that all traces of our late unhappy civil strife may be speedily removed. These ends can be easily reached by acquiescence in the results of the conflict, now written in our Constitution, and by the due and proper enforcement of equal, just, and impartial laws in every part of our country.

Grant reflected the Radical Republican strategic assumption that the issues of the war had been settled and that the South must accept the new conditions as written into the Constitution. The president made an attempt to gain some advantage on the information battlefield by placing much of the federal response on the local actors. "The failure of local communities to furnish such means for the attainment of results so earnestly desired," he concluded, "imposes upon the National Government the duty of putting forth all its energies for the protection of its citizens of every race and color, and for the restoration of peace and order throughout the entire country."[4]

The War Department message carried the secretary of war's endorsement, and, in accordance with the president's direction, all regular forces of the United States stationed in any locality were ordered to take action "whenever the occasion shall arise" against perpetrators of the crimes outlined in the April 1871 act by "assisting the authorized civil authorities of the United States in making arrests, preventing the rescue of persons arrested under the law, in breaking up and dispersing bands of disguised marauders, and of armed organizations."[5]

Congress had responded to the asymmetric challenge of white Southerners with a high-risk approach that was intended to intimidate individuals to prevent them from approaching the mobilization threshold. It was a signal that federal power could now fall directly on individual citizens at will, with no guarantee of protection from arbitrary arrest and imprisonment. Presented as an act of righteous protection of basic rights guaranteed under the Fourteenth and Fifteenth Amendments, Congress was attempting to reassert the moral authority of 1867 and regain the high ground on the information battlefield. The danger was that such dramatic action would fail to bring about any significant change. Military force would be used largely symbolically to support civil authorities, rather than as a stand-alone force intended to locate and destroy armed groups. Such a course of action would almost guarantee an open war.

The dilemma the president and Congress faced in attempting to deal with this asymmetric approach was cogently summarized by Job Stevenson: "For us to submit would be ruin; to resist would be war, and if it comes to war, then we shall see in this pleasant land the horrid front erect of mediaeval, savage, barbarous war." The dominant actor had the power available to overwhelm any armed resistance but would not accept the political cost such an action would entail; nor were the Republicans willing to take the risk of returning to open warfare. Thus, a political solution had to be found. The deployment of army units in the South from 1870 to 1876 became largely a symbolic deterrence measure intended only to demonstrate support to the Republican state governments, not the will or authority to carry out any decisive action.

In many parts of the country, the news of the Ku Klux Klan Act and Grant's proclamation was met with skepticism. These reactions had a certain restraining effect on Congressional Republicans' intent to reverse the increasingly troubling conditions in the newly reconstructed states. In far-away Idaho territory, the government's actions were seen as nothing more than the Radical Republicans in Congress giving the president extraordinary powers to use military force in the South as a cynical political ploy to ensure his own reelection in 1872. Similar doubts were expressed in other western newspapers. One of the most damning was a commentary in a Carson City, Nevada, newspaper, which described Congressional alarm over the Ku-Klux as "exaggerated, unfounded unwarranted senseless sensational humbug." It went on to proclaim:

The Ku Klux Klan has been made the pretext for nearly every outrage that has been perpetrated upon the Southern people. . . . and in nearly every instance where any of these outrages charged to the Ku Klux Klan has been investigated, or where an investigation could be reached, the charge has fallen to the ground.

The *States and Union* of Ashland, Ohio, was dismissive of the extent of the threat represented by the Ku-Klux and suspected a political agenda, as did the *Evening Telegraph* of Philadelphia, which expressed frustration with the continuous meddling of Congress in the Southern states. In general, those newspapers hostile to the efforts of Congress indicated a suspicion that the reason army units were being dispatched everywhere in the South was not to suppress Klan violence, but to protect the discredited Republican state governments.[6]

Following the pattern it had established earlier, Congress held hearings to establish its position and set conditions for further action. The Joint Select Committee to Inquire into the Condition of Affairs in the Late Insurrectionary States, as it was called, was composed of 7 senators and 14 congressmen. Dividing into subcommittees, they visited South Carolina, North Carolina, Georgia, Florida, Alabama, and Tennessee, and they held hearings to assess the scope and scale of the resistance in those states. Between May and September of 1871, the committee assembled a body of testimony that, even today, is numbing in its portrayal of fear and suffering. But the testimony of white Southerners also provides an insight into the conditions that led to a mobilized identity to foster and sustain an asymmetric resistance that the committee completely overlooked.[7]

Augustus R. Wright, a judge and former member of both the U.S. and Confederate Congresses, who had opposed secession but served as a commander of a Georgia infantry regiment, provided in his testimony a powerful and insightful expression of the mobilized collective identity and how Ku-Klux violence had its own purpose and logic. He made it clear to the committee members that however wrong the actions of the Ku-Klux were, these actions served a purpose. The Ku-Klux's purpose, Wright explained, was "to conserve what they take to be the life of society," which they believed was "seriously endangered." As long as this fear existed, Wright said, "I do not think a million of troops would prevent it." This description of "the life of society" can be understood as a defense of the collective identity against a threat to its existence and the willingness to take drastic action to defend it, even in the face of overwhelming force. John B. Gordon, one of the most famous generals in the Confederate army, and at the time of his testimony, a Georgia businessman (and the reputed leader of the Ku-Klux in Georgia), testified that the Ku-Klux in Georgia was made up of "peaceable, law abiding citizens," who had bonded together for self-protection against the efforts of the carpetbaggers to organize the freedmen into quasi-military political organizations. Gordon described the Ku-Klux phenomenon as a reflection of what he called "apprehension" that "took possession of the public mind of the State."[8]

Congress was not interested in the sources of the resistance of the weaker actor; instead, it had to challenge the growing advantage of the asymmetric challenge to the dominant actor. The committee had a larger political-strategic purpose and needed consistent messaging from those testifying to build a public image that would stir action in the North and recapture the momentum of their failing strategy. To do so, the committee sought to build a mass of evidence to justify the Ku Klux Klan Act. Therefore, it was necessary that the testimony given before the committee was properly ordered and regulated to serve as a counter-narrative to the Southern resistance. The testimonies were shaped to repeat the consistent theme throughout the South of the weakness of the state governments and the vulnerability of the freedmen to highlight the need for federal intervention. It was important to reinforce the image of the Ku-Klux for a Northern audience that was already well acquainted with the image of the Klan as a massive, all-pervasive organized force instigating chaos and terror. In the pattern of their questioning, the Republican committee members consistently prompted witnesses to attribute all incidents of violence (where violence was a part of life in the postwar South) to the Ku-Klux.[9]

When the Congressional subcommittee visited South Carolina in the summer of 1871, one local newspaper reported that the committee had interviewed 100 witnesses at Columbia, Spartanburg, and Yorkville, 53 of whom were freedmen. The newspaper scornfully noted that the testimonies given by these witnesses were characterized by a "paucity of facts," and observed that issues of labor disputes and violations of social caste were the source of most violent activities, not politically motivated violence. The *Daily Phoenix* in Columbia reported with undisguised disdain that government witnesses were secured in the county jail until they gave their testimony, then were released and paid.[10]

The Congressional counteroffensive got off to a somewhat inauspicious start. A *New York Herald* correspondent traveled to South Carolina in September to assess the local conditions in the state and encountered a federal official who told the newspaperman that he believed "the Ku Klux Klan was the most perfectly organized conspiracy in the country." While a number of "diabolical outrages" had been committed by "brutal hardened scoundrels," the number of individuals actually arrested for Ku-Klux-associated crimes were mostly small farmers who claimed they had organized themselves under the Klan name for mutual protection. The *Herald* reporter subsequently visited 40 men detained in Spartanburg, South Carolina, under the Ku Klux Klan Act and noted that many had surrendered voluntarily after learning of President Grant's proclamation. He also found that at least half of the men would not be charged with any crime and released. Another federal official told the reporter that "the government must prove its power," but there had been no acts or resistance in more than two months, so there had been no need to suspend habeas corpus because there had been no resistance to the execution of federal arrest warrants.[11]

The burden of effort to demonstrate the effectiveness of the federal government to enforce its will under the Enforcement Acts fell to the attorney general, Amos T. Akerman. Akerman certainly understood that the government was confronting an unfamiliar and extraordinary situation in the South. He characterized it as war and, therefore, believed it required a warlike response. Unfortunately, law-enforcement officials (U.S. marshals and federal attorneys), even though armed with the ability of arbitrary arrest and detention without evidence or a specific charge, still needed the support of the army. At this time, there were about 8,000 troops in the Southern states, nearly one-third of them guarding the border in Texas. General Terry, the commander of the Department of the South (Tennessee, Alabama, Florida, North Carolina, and South Carolina), made it clear that the warlike measures Akerman envisioned—a massive deployment of force throughout the South operating in partnership with federal officials—was impossible. The entire U.S. army was far too small to conduct such a mission, the state governments were too weak to provide any substantial assistance, and the number of federal law-enforcement officials was insufficient. Therefore, the only other course of action that Terry could support with the forces he had available was far more modest and heavily dependent on the effect it would produce for success. This involved blanketing a targeted area in one state with troops and federal law enforcement officials in an effort to eliminate resistance, with the intent of demonstrating a level of federal strength and resolve that would serve to deter resistance in other states.[12]

Akerman traveled to South Carolina to meet with Maj. Lewis Merrill, the U.S. forces commander in the state, who had over time developed a network of informers in York County and was confident that he had identified the resistance members in the up-country. He reported that "the object of the organization is to terrify the negroes into obeying the whites in voting or to compel them to stay away from the polls." He related that there had been between 300 and 400 incidents of violence reported between November 1870 and July 1871.

In 1870, the Democrats in South Carolina renamed themselves the Union Reform Party and nominated R. B. Carpenter, a circuit judge who came to South Carolina from Kentucky in 1867. The candidate for lieutenant governor was Matthew C. Butler, who had served in Hampton's Legion, the famous South Carolina cavalry unit. He had entered the war as a captain, been wounded twice, and emerged as a brigadier general. This alliance of carpetbagger and Confederate indicated that a new political strategy was being implemented on the example of Virginia, which had enjoyed little interference with military occupation or political turmoil by building political alliances between factions and gaining enough votes to be elected. Butler and Carpenter even went as far as to hold a barbeque in Edgefield to celebrate the adoption of the Fifteenth Amendment. Butler called the amendment "the climax of Reconstruction" and expressed great joy that "the colored man had been set free." Carpenter described the amendment as "the grand culmination of the war," representing "the statutory pacification of the country." Carpenter explained the barbeque as a way to celebrate the event that gave blacks "the blessings of civil and political liberty."[13] Despite these appeals to South Carolina's black voting majority, through celebratory barbeques, promises of reform, and appeals to trust former slaveholders, nothing was sufficient to overcome Republican Party loyalty, which was bolstered by the Union Leagues and the existence of a state militia with 14 black regiments.[14]

In fact, the South Carolina version of the Klan operated after the election of 1870 in a different form, that of a paramilitary organization to counter the state militia. Raids were conducted to seize weapons and ammunition; individual militia members, Union League members, Republican officeholders, and voters were targeted with the intention to neutralize their political participation. Interestingly, those Republicans the Klan deemed as honest and competent were not subjected to violence. Violence was the means, not the end, representing a resistance to what the majority of whites believed to be the injustice, tyranny, and corruption of the state government directed against them. By 1871, the resistance was operating with impunity in York County, essentially acting as the de facto enforcers of rough justice. Putting on some form of recognizable Ku-Klux garb became the symbol of taking action—a crossing of the mobilization threshold. Ku-Klux violence was localized but so widespread as to appear pervasive and organized. Often spontaneous, it was intended to enforce a social order between whites and blacks through vigilante discipline and punishment where no faith in law and order existed. The lawless conditions also encouraged opportunities for criminal mischief as well as politically served violence and intimidation.

Criminal activity became more brazen, including a military-style raid of 100 men on the county treasurer's office. A South Carolina newspaper, reporting on the increased level of activity in the up-country, refrained from endorsing or supporting their efforts but did acknowledge that the Ku-Klux represented "men of firmness and nerve, who strike because they believe it necessary for the protection of their life, property, and liberty." The article expressed the belief that the Ku-Klux arose from ruinous state activities, ranging from election fraud, to arming freedmen and forming a militia for the suppression of white Southerners, to enormous taxation. If the state would practice good government, employ prudent expenditure of funds, disband the militia, and appoint intelligent jurors, the Ku-Klux would disappear.[15]

Convinced that General Terry's course of action was the only possible path to regaining the initiative, President Grant issued a two proclamations in October. The first was a warning requiring all illegal combinations and conspiracies to disperse and surrender arms, ammunition, and disguises to federal authorities within a five-day period. The second proclamation followed after the expiration of the warning period, and activated the Ku Klux Klan Act in nine counties in South Carolina, suspending the writ of habeas corpus and authorizing federal intervention. At the same time, the redoubtable Senator Morton gave a speech in Indianapolis intended to galvanize Northern support for reinvigorating the Congressional Reconstruction strategy by summarizing some of the evidence gathered through testimony to the Congressional investigating committee on the Ku-Klux, cataloging a series of "outrages," including murder, intimidation, whipping, and other crimes (e.g., rape, arson, robbery, and forcible depopulation).[16]

A company of U.S. infantry was located in Yorkville and had been reinforced by an additional infantry company and two cavalry companies in 1871. Within 48 hours of the War Department's authorization for military action with the president's suspension of habeas corpus, these units swept through the countryside, initially arresting 82 men. Another 500 turned themselves in, and several hundred fled the county. The government forces took to the task with alacrity, conducting nighttime raids and pursuing fleeing suspects. Between October 1871 and January 1872, forces detained 472 men; three months later, the total of number of individuals detained had reached 533. All in all, more than 1,500 would be caught up in the sweeps. Maj. Lewis Merrill, the federal force commander, expressed his frustration, noting that despite what appeared to be a significant victory against the resistance, "the very worst of these men if released tomorrow would be received . . . as heroes whose sufferings had been martyrdom for the rights of the South."[17] Merrill recognized the significance of the mobilized collective identity that would not be influenced by military action, but instead would be even more strongly unified in protecting and defending that identity from a now very visible threat.

The Klan trials held in South Carolina over two court sessions were sensational events, with the defendants represented by a very powerful legal team supported by contributions from all over the South, another indication of the mobilized collective identity. The presiding judge, Hugh L. Bond, reflected the deep prejudice of Northerners against upper-class whites, whom they believed were responsible for secession and misleading the Southern yeomen and poor whites to a ruinous war, and who now were replicating their infamous and traitorous behavior by leading underclass whites in the Ku-Klux outrages. Thus, every upper-class white man who faced trial was given a stiff sentence, and lower-class whites were treated with leniency. These sentences may have had a strategic intent in attempting to divide Southern whites and both weaken the leadership of and fragment the resistance.[18]

Out of the thousands detained, only a handful actually stood trial, one of whom simply walked out of the courtroom during a recess and disappeared, much to the embarrassment and consternation of the authorities. A total of 37 sentences were handed out by juries composed of black citizens and one white Republican. Those brought before the court were not allowed to testify during their trials, but they were allowed to make a statement before sentencing. This tactic was intended to give the defendant the opportunity to make a public confession and demonstrate remorse so that the government, in return, could demonstrate its stern effectiveness. This did not work to any effect, leaving Judge Bond to make awkward pronouncements favoring farmers who could not afford either to pay fines or face a prison term, which would leave their crops in the field and their families without support.[19]

The government's effort to demonstrate its power was seriously impaired when Akerman resigned and George H. Williams became attorney general in 1871. Williams was a less than enthusiastic supporter of the rights of the freedmen, and he had little interest in the ongoing federal enforcement activities in South Carolina. Williams quickly discovered upon taking office that the Justice Department had run out of funds and could not continue the trials. Instead, he stressed as much as possible the positive outcomes, asserting that the trials had had the desired effect of bringing about public peace and deterring future outbreaks of violence. Williams's backtracking was a serious self-inflicted wound that negated any advantage the government may have gained earlier on the information battlefield. Congress and the president had proclaimed a willingness to take extraordinary actions in defense of the Constitution and the Congressional strategy for Reconstruction, but as a dominant actor, those actions belied a lack of will to achieve its strategic goals. Indeed, almost forgotten in the flurry of action of the summer of 1872, was the act of Congress in May known as the General Amnesty Act, which terminated all political disabilities on certain specified former Confederate leaders from holding office, except senators and representatives of two prewar Congresses; judicial, military, and naval officers; as well as heads of departments and diplomats. Although a minor modification in the overall political landscape and largely symbolic, it nevertheless was a significant indicator that the Congressional strategy was in danger. By 1873, Arkansas, Louisiana, Missouri, and Virginia had repealed restrictions on the franchise; with Democrats taking control of Alabama and Tennessee, enfranchisement restrictions were changed significantly. Disfranchisement of influential leaders had been a significant element of the strategy to support the survival of the Republican governments in the

South and to demonstrate the power of the dominant actor. The Amnesty Act opened the door for a resurgent Southern leadership to support a new dimension of asymmetric warfare.[20]

Although the government claimed victory in South Carolina, and despite the efforts of the most vociferous Radicals in Congress to highlight the Congressional report on Ku-Klux violence to sustain Northern will, and even the subsequent claims of historians that a vigorous federal response broke the back of the Klan in South Carolina, the actual results were tepid at best. The voluntary confessions and surrenders of purported Ku-Klux members were indications not of a dedicated guerrilla organization with solidarity, cohesion, and discipline but of a mobilized identity with varying levels of commitment. Many of those who turned themselves in were afraid of punishment or experiencing individual remorse for their actions. The mighty hammer blow intended to fall on an imagined armed guerrilla force instead fell on thin air. The violence in South Carolina represented the last real vestige of Klan activity in the South. The actions had served a larger purpose as part of asymmetric warfare. It had already largely passed elsewhere as changing political conditions established a new threshold of action. A special correspondent from the *New York Tribune* reporting on conditions in the South in August 1872 came to a conclusion that appeared to elude the federal government. "I am convinced," he wrote, "that the Ku-Klux spirit is now dead."[21]

The Grant administration's attempt to regain the initiative had failed. In an article published in December 1872, a *New York Herald* correspondent traveling through the Gulf States observed that "the great curse of society there has been politics," and concluded that "Reconstruction has been a bungle in the South." The Ku-Klux, he noted drily, "was the first product of political ferment." He found the post-Republican governments of Texas and Alabama the most stable, working, he told his readers, to "palliate, counteract and provide for the excesses of carpet-baggery."[22]

In fact, prosecutions under the Ku Klux Klan Act ended in 1873, and shortly afterward, President Grant issued a proclamation of clemency to those still awaiting trial and to those who had been convicted under the act. Quite simply, as one historian has concluded, "the enforcement efforts of the national government failed to change the hearts and minds of white South Carolinians." Indeed, these enforcement efforts failed to influence the course of events in South Carolina and elsewhere as Southerners continued to advance their asymmetric advantage against the Republican state governments.[23]

The Ku Klux Klan Act and the trials actually marked the turning point to Phase III of asymmetric warfare in the South. The Ku-Klux movement did indeed dissipate, presenting a false image of federal victory; but, in reality, it only marked the point where the mobilized Southern identity had passed the decisive threshold of support. White Southerners recognized that the will of the national government was weakening as the main instrument of enforcement, touted as the decisive effort in destroying Southern resistance to Congressional Reconstruction, had been nullified in 1873 by the president and the Justice Department.

Even as the federal government was mobilizing resources to suppress the Klan in order to regain control of its Reconstruction strategy, the Ku-Klux phenomenon had run its course. It was already in the process of changing form, from spectral night riders to paramilitary organizations overtly supporting Democratic Party candidates. Local armed elements operating with military efficiency and discipline also emerged, demonstrating a willingness to move to extremes if provoked. The ever-present possibility of an escalation to a race war was something unthinkable to the Republican leadership and a significant limiting factor in responding to the resistance. Here white Southerners demonstrated an asymmetry of will in adapting the type and means of resistance to existing and emerging conditions. The Klan had served its purpose. It would be replaced by a different entity with a similar, but far more overt political purpose.

None of this could have happened without a significant shift in Northern public opinion against the Grant administration's taint of corruption and against the Republican state governments, now nearly universally identified with carpetbag exploitation and corruption and black misrule, maintained exclusively by military force provided by the federal government. This white Southern representation of conditions continuously shaped and influenced the national will to continue Reconstruction as it had been formulated by Congress. As the dominant actor's faith in the transformation of the South under the supervision of Congress began to fade, the Republican state governments became increasingly vulnerable.

General Terry's recommended course of action for a concentrated federal effort in South Carolina as a demonstration of force could only be successful if it served to deter or dissuade the resistance in other states from continuing their activities. From this perspective, the South Carolina operation was a total failure. For even as South Carolina was being used to demonstrate federal resolve, North Carolina was slipping away from the Republican sphere of control.

7

Asymmetric Warfare Phase III:
The Dominant Actor Dislocated
(1873–1876)

Asymmetric warfare depends on a continuation of violence that serves to keep the dominant actor at a disadvantage to the weaker actor. If the dominant actor cannot regain the initiative by protecting its vulnerabilities, countering the level of violence, and competing on the information battlefield, the weaker actor uses violence to support larger ends that mark the initiation of Phase III. In this phase, electoral politics and violence (or the threat of violence) are overtly combined for the strategic purpose of shaping political outcomes in favor of the weaker actor, while simultaneously continuing to influence morale and support for the dominant actor on the information battlefield. In Phase III, the mobilization threshold within the Southern-Confederate collective identity was crossed, as more individuals actively participate in the belief that the benefits of participation far outweighed passive support. The goal in Phase III is to continue to limit the dominant actor's options, gain the initiative through dominance of the information battlefield, and transition the mobilized collective identity into a popular movement. As these goals are achieved, violence in Phase III becomes less important, and armed groups emerge as overt paramilitary organizations clearly allied to a political organization manifesting itself as a popular movement seeking to supplant the dominant actor's influence and power.

Phase III in the Reconstruction South marked the shift in the distribution of power from the freedmen back to white Southerners. Asymmetric warfare's Phase III emerged as the prospect of effective federal intervention diminished, political conditions increasingly favored the Democrats, and as the Southern representation of events came to dominate the information battlefield. For some states, such as Tennessee, North Carolina, and Georgia, the transition from Phase II to Phase III was rapid; for Texas and Mississippi, the transition to Phase III was marked by a steady level of violence demonstrating an asymmetry of will that the Republican governments were unwilling to match. For South Carolina and Louisiana, Phase III included a high level of violence and open demonstrations of military power by armed groups to build sufficient political momentum to overcome the Republican governments. The freedmen were gradually neutralized as political actors and as a source of power of the Republican state governments through violence, intimidation, fraud at the polls, open armed demonstrations, and, in some cases, open combat. In the process, white Southerners gained the initiative to move to a political strategy of creating a spontaneous popular movement aimed at gaining control of the state governments. Within this popular movement, the Southern-Confederate mobilized collective identity demonstrated a powerful capability in which a desire for freedom and democracy easily existed side by side with racial superiority.

William W. Holden was inaugurated as the governor of the newly reconstructed state of North Carolina in July of 1868. Holden was a newspaper editor and politically active Democrat before and after the war. In 1864, Holden ran as the peace candidate for governor of North Carolina. In a remarkable parallel to future Reconstruction events, armed men guarded the ballot boxes designated for Holden to threaten and intimidate Holden supporters and suppress their votes. Facing numerous death threats, Holden went into hiding, fearing an assassination attempt before the votes were counted.

After the war, President Johnson appointed him as the provisional governor, pending the establishment of a new state government under the conditions of Presidential Reconstruction. With the invalidation of the North Carolina state government under Congressional Reconstruction, Holden moved to the Radical Republican camp and was elected under the new state constitution mandated by Congress. In his inaugural address, Holden had nothing but conciliatory words to offer, and it appeared that he was attempting to forestall what he undoubtedly sensed, given his previous experience in state politics, was potential trouble ahead. While he asserted that the new government was "for the whole people," he also was quick to add that the government was "not enemies of any portion of the people of the State," and expressed with hopeful determination that the government was "friends to all."[1]

The governor certainly understood that he would be heavily dependent on the goodwill of a powerful and unified majority that had little interest in cooperating, no matter how soothing the words he had spoken. Many former Confederates remembering 1864 considered Holden a traitor to the cause. Moreover, Holden's grip on power was dependent on a weak and factionalized Republican Party, consisting of an uneasy alliance between mostly Unionist whites in the western part of the state and the freedmen.[2]

At the time of Holden's inauguration, there were more than 900 U.S. soldiers in the state, divided among three garrisons. A year later, that number had declined to 366. These soldiers could only be dispatched upon authorization from the War Office in Washington and only in the case of open resistance or specific emergencies. In North Carolina, armed groups representing numerous interests had many different names: Heroes of America, Loyal Union League, Red Strings, Constitutional Union Guard, White Brotherhood, Invisible Militia, North Carolina State Troops, North Carolina Militia, and Jay Hawkers, as well as the Ku-Klux. These groups not only competed with one another but also battled with state authorities. In the absence of state power, some groups took the law into their own hands; some groups became lawless bands of marauders, and others directed violence against the state. The violent actions of these groups combined to create an effect of unrest that contributed to the asymmetric actor's goal to delegitimize the Holden Republican government. Although a state militia was organized in 1868 to provide security and order, few whites joined. Without white support of the militia, Holden's ability to employ a security force was extremely limited. A militia force was only effective if it was seen as nonpolitical and had broad representation of state citizens. Like every other organization related to the Republican state government, the militia was composed of only loyal Republicans and was, therefore, viewed as a partisan political instrument of force serving only to keep the party in power by targeting former Confederates. This image was dangerously compounded with black militia units. Rather than contributing to security, they presented the great risk of enflaming an already dangerous situation. The presence of armed blacks in the midst of an already restive white population suspicious of the motives of the governor and fearful of black depredations under state authority and sanction, would generate antagonisms and even greater violence, perhaps igniting a race war. Although Holden had the power to establish order, he hardly dared to use it.

During the 1868 presidential election in North Carolina, the Ku-Klux attributed resistance was strongest where the balance of political power was the most tenuous. Where black majorities existed, there was less Ku-Klux-related activity. Rockingham and Alamance Counties were particularly active centers of resistance.[3] The Raleigh *Weekly Standard* bravely asserted that "the Union men of the South can take care of themselves, if only

Congress will sustain the President in taking care of their enemies." Taking care of enemies, according to the editors, meant suspending the writ of habeas corpus, holding military tribunals, and executing Ku-Klux members.[4] Such a pronouncement only indicated the extreme weakness of Holden's Republican government. Because the state constitution forbade the suspension of habeas corpus and military tribunals, the Republicans were signaling that without federal intervention, martial law, and summary executions, they were otherwise helpless before the resistance.

Between 1868 and 1870 there were 78 instances of Ku-Klux-related violence, including the murder of freedman Wyatt Outlaw in February 1870. Outlaw was a civil magistrate and also president of the local Union League. He had been identified firing at a group of Ku-Klux on parade. Holden responded with an appeal to the army, which sent a detachment of 40 men to the county. In May 1870, John W. Stephens, a Republican state senator from Caswell County, was murdered. Stephens, who spent the war as an agent of the Confederate government confiscating horses, had built up a great deal of resentment against him. After the war, he became a Freedmen's Bureau agent and, as a loyal Republican, organized a Union League and made a number of incendiary speeches. Popular among the freedmen, he was elected as a state senator but was widely ostracized and held in contempt by the white population. His death was unlamented, and, in fact, the murder was not even reported to the governor after the discovery of the body.[5]

In January 1870, the state legislature passed a law authorizing the governor to declare a state of insurrection and deploy the militia to where the Ku-Klux was most active when local authorities could no longer provide law and order. In the counties where the Ku-Klux was most active, the white population had passed the threshold of support and had become mobilized, actively supporting the resistance and establishing what amounted to ungoverned space. The Ku-Klux became better known as the White Brotherhood, whose object was to overthrow the Republican Party and advance the conservative party. It was estimated that 10 camps existed in Alamance County with a total of 700–800 men; Guilford County was estimated to have 1,200 men, and a total of 40,000 were thought to be operating throughout the entire state.[6]

In July 1870, President Grant sent federal troops to Alamance and Caswell Counties, the heart of the Ku-Klux resistance, but the troops had no effect. Restricted by orders to avoid confrontations and act only in extraordinary circumstances when called upon by state or local authorities, the soldiers remained passive observers. By September, Governor Holden had placed Alamance and Caswell Counties under martial law. Unwilling to rely on black militia and disappointed with the results of a white militia force, Holden sought the services of a professional—Col. George W. Kirk—a former Union officer who during the war had fought Confederate guerrillas with the Second and Third Mounted Infantry Regiments in a bitter campaign in both western North Carolina and east Tennessee. Kirk's units were composed of independent mountaineers with no interest in slavery who were natural allies of the Republicans. After the war, Kirk had been an officer in the Tennessee State Guard and had led a number of operations against Ku-Klux resistance. Kirk brought with him a collection of seasoned fighters who would serve the governor's purpose. Given arbitrary arrest powers and backed by a military court, Kirk dispatched his forces on raids and arrested about 100 individuals. Several of these captured individuals made a public confession printed in the pro-Republican newspaper in Raleigh. The White Brotherhood, they claimed, was intended "to defeat entirely the reconstruction acts and to deprive the negro of all privileges as a citizen in the county." Each individual renounced the organization "as dangerous to civil liberty, as calculated to tear down a Republican form of government and finally to involve our whole country in civil war and bloodshed."[7]

Most of the individuals rounded up were later released for lack of evidence by a federal judge, who found that the governor had violated the state constitution's prohibitions against declaration of martial law and suspension of the writ of habeas corpus. Other detained individuals were taken to the state capital at Raleigh, where three state supreme court justices began hearing testimony related to whippings, beatings, torture, and assault that had been attributed to the Ku-Klux. About 49 were held for future trial and released on bond. Thomas Settle Jr. was one of those justices. After hearing these accounts, he had no doubt that the security situation in the state was rapidly deteriorating. It appeared to him at the time that the Confederacy and rebellion "were about to triumph." Although Settle attributed the violence to "the hostility and bitterness growing out of the war," he never recognized that something far deeper and more significant was happening.[8]

A Washington, D.C., newspaper related the reporting of a *New York Tribune* correspondent traveling in North Carolina to examine the conditions related to the Ku-Klux movement. The correspondent interviewed W. A. Smith, a Republican and railroad executive, who revealed the depth of Southern opposition to the authors and agents of Reconstruction. Smith said Northerners had no conception of the level of hatred and bitterness that was directed at the federal government. The state government and Negro equality, he added, are equally hated. Smith gave a bloodcurdling account of the Klan's activities in 10 counties of North Carolina and predicted that in the upcoming elections, Republicans throughout the entire South would soon experience Southern hostility and, as he put it, "the whole power of their Ku Klux hell-hounds."[9]

Although Holden declared victory over the Klan, his extralegal actions proved to be a great embarrassment to the Grant administration, which at the same time was employing forces against the Klan in South Carolina. The collapse of the effort to suppress the violent Ku-Klux resistance in North Carolina proved to be the last effort of the Republican state government. Holden's deep unpopularity with white Southerners based on his wartime betrayal as a peace candidate and his subsequent turn to the Republicans prevented any opportunity for support. The layered violence that had plagued North Carolina since the war had an additional layer placed on it with the advent of the Ku-Klux in 1868. The levels of violence served the political interests of the mobilized Southern identity, which sustained the resistance, defying the state authorities until Holden launched a last-ditch effort to overcome the Ku-Klux organization. But the violence subsided with military deployment, and sensing that the political winds were changing, the Klan and the White Brotherhood faded away, although violence in some areas continued into 1871.[10]

Support for the Republicans collapsed, and the Democrats gained control of the general assembly in the 1870 elections. Holden was plagued by lawsuits brought against him by individuals detained during the militia sweeps. The state legislature finally indicted and impeached him for abusing his authority, violating state law, and misusing the power of his office. Turning his duties over to the lieutenant governor in December 1870, Holden was found guilty in March 1871 and removed from office. The *Charlotte Democrat* observed that "Holden and his official supporters have failed to maintain themselves in any way, foul as well as fair, in their State; they have appealed to popular elections, and have been rejected."[11]

Colonel Kirk and his militiamen wisely left the state. Shortly thereafter, the legislature issued pardons for all crimes committed by secret organizations, including Union Leagues. In addition, the law giving the governor power to declare insurrection and the law authorizing the governor to organize a state militia were repealed. Republican strongholds in the state were diluted through redistricting to avoid the possibility of a resurgence of Republican power. In 1873, an amnesty act was passed that pardoned anyone "who may have committed any crime against or violation of the laws of the State."[12]

North Carolina experienced the three phases of asymmetric warfare in rapid succession. After the mobilization of the collective identity of white Southerners, came the Ku-Klux and the direction of a level of violence against political targets to weaken and delegitimize the Republican government. Led by an unpopular governor and plagued by a weak governance structure, the state was vulnerable. The layered violence that enhanced the progress of asymmetric warfare succeeded in pushing Governor Holden to take a great political risk in the hopes that results would justify the methods. His political miscalculation to deal with the violence led the Democrats to gain control of the legislature and offer a level of control to white

Southerners. The governors of Arkansas and Georgia experienced similar political miscalculations, which led to a series of political counterstrokes that ended Republican control in their states.

As intended by the Congressional Reconstruction strategy, the votes of the enfranchised freedmen in the 1872 election ensured Republican Party dominance and returned President Grant to the White House. South Carolina had the highest percentage of Republican votes than any other state in the Union; Mississippi was close with 63 percent of its votes cast for Republicans, at about the same percentage as Pennsylvania and Michigan; Louisiana had over 55 percent of votes cast for Republicans, which was more than Ohio, and similar to the percentages cast by voters in Illinois and New Hampshire. But there were clear indications that the dominant actor's will to carry out the Reconstruction strategy was faltering. Several key Republican senators who had been the leaders of the strategy's formulation and implementation, Carl Schurz, Lyman Trumbull, and Charles Sumner, had joined a liberal Republican faction that would become a strong opponent to President Grant's efforts to sustain Reconstruction. The faction desired to turn the party toward efficiency and honesty in government and reduce expenditures. Even as the Grant administration was attempting to suppress the Ku-Klux in South Carolina, Schurz called the Enforcement Act unconstitutional and a violation of states' rights. He called for an end to the Republican Reconstruction governments and return of control to white Southerners, who he believed were the only group able to establish peace and order.[13]

Not surprisingly, Southerners strongly supported the message of the liberal Republicans and would use it to their advantage in the coming years on the information battlefield. After the election, the Grant administration was battered by the exposure of a number of scandals that associated Northern Republicans with the same kind of graft, greed, criminal activity, and corruption that Southerners had long associated with the Republican state governments. One of the biggest scandals was the revelation that the Republican government of Louisiana was supported by a network of corruption led by one of the president's in-laws.[14]

By 1874, the combined factors of pervasive Republican corruption and a significant economic depression changed the environment in favor of the weaker actor. The state governments of South Carolina, Mississippi, and Louisiana, already under siege, were now facing economic ruin. Taxes, already high, were forcing land confiscation for nonpayment of taxes as the price of cotton steadily declined. Hard economic times only served to focus greater resentment against the Republican governments. Southerners now had a different focus of their efforts to delegitimize the state governments. Their attacks focused on the corruption of power. The Republican state governments had come to power unchallenged and secure through the votes of the freedmen. As a result, Southerners claimed, these governments misused power and became irremediably corrupt. The injustice of marginalization and the oppression of white Southerners who had been forced to endure this new order for an intolerable length of time became the watchwords of the mobilized collective identity to shift to an all-out effort to change the status quo. The compelling need for relief required the removal of Republican control and the restoration of honest government. As the third phase of asymmetric warfare changed the political landscape of the majority of Southern states, the mobilization of whites in South Carolina, Mississippi, and Louisiana manifested itself as an open political-military challenge to the Republican governments. This mobilized collective resistance represented a unique adaptation to the conditions that existed in these states.

Whites assembled as taxpayers and good government advocates who were willing to resort to open armed resistance directed at the security forces of the state to eliminate their ability to protect blacks as the primary political actor and base of Republican power. Defeating state security forces and neutralizing black votes would open the path for political success through the employment of a paramilitary arm paired with Democratic Party candidates. This open challenge to the Congressional Reconstruction strategy represented the weaker actor's recognition that the dominant actor had few options available to respond and was no longer capable of mustering the will necessary to respond effectively. This asymmetry of will employed by the weaker actor was demonstrated in a most dramatic fashion by the willingness of the Southern paramilitary forces to initiate a race war, a situation that was by no means acceptable to the dominant actor. This asymmetry of will led to control of both the physical space within the state and the political actors. In addition, the weaker actor achieved complete dominance of the information battlefield as more and more Northern newspapers reflected a willingness to make some accommodation, while accepting the Southern prescribed image of the cynical carpetbagger exploiter ruining the states.

In Louisiana, fraud and violence during elections was commonplace. The state had been under military occupation since 1862 and had been functioning under a constitution ratified by loyal whites in 1864. Henry Clay Warmoth, a former Union lieutenant colonel from Illinois, became the governor of Louisiana at the age of 26 when the state was readmitted to the Union in 1868. With the onset of Congressional Reconstruction and a new constitution, Republican leaders believed that educated free blacks and ambitious former slaves would become the core of the new party, and they would be the leaders of the black majority that would deliver the votes and, in partnership with loyal whites, assure Republican control for decades. However, this formula failed as loyal whites were just as opposed to unlimited suffrage extended to freedmen as the white majority.[15]

Warmoth was a political chameleon. A carpetbagger and bête noire of the resistance in Louisiana, he faced impeachment charges in 1872 for official misconduct as governor of Louisiana, but remained active in state politics, serving in the legislature from 1876–1878. He ran unsuccessfully for governor in 1888, and from 1889 to 1893, he was the collector of customs for the Port of New Orleans. At first a strong supporter of black suffrage and civil rights, he shifted continuously as the political winds moved in favor of allying with whites, but he ensured Republican control through the establishment of the state returning board to supervise election returns. Staffed with reliable political allies, the board ensured that Republican candidates would be certified as the winners. In addition, the state legislature established a state militia, under the command of James Longstreet as adjutant general. Longstreet, a corps commander in Lee's Army of Northern Virginia, was one of the heroes of the Confederacy. He had relocated to New Orleans after the war and joined the Republicans. This act of apostasy to the Southern-Confederate collective identity earned him scorn and repugnance. Moreover, the militia he commanded was made up almost entirely of blacks.

The rise of a Phase II organized resistance in Louisiana was more politically motivated than in other states and was not always associated with either the Ku-Klux or the Knights of the White Camellia, which was the more common name of armed groups in Louisiana. Examples of the effects of this more direct violence are clearly seen in a number of depositions made in various parishes in Louisiana in 1868. At Rapides Parish in April on election day, a U.S. army sergeant of the 20th Infantry reportedly stood at the door of the courthouse, took Republican ballots away from freedmen, and replaced them with Democratic ballots. A group of disguised and well-armed white men (because of disguises they were automatically associated with the Ku-Klux) paraded at night, along with 50 policemen appointed by the mayor (among them boys under the age of 14, three of whom were black), confiscated weapons from freedmen coming to town to vote. The weapons were never returned.[16]

A letter to Governor Warmoth from Madison Parish in August reported that "considerable excitement prevails in consequence of the many reports of the murder of white Union men in all the border parishes." The letter writer asserted that all other Republicans were being repeatedly threatened and that "the old citizens of this parish," as he described them, were "bitterly opposed to the present State government."[17]

During September and October of 1868, in Orleans Parish, pro-Republican processions were assaulted and fired at, and threats were made against anyone who identified himself as a Republican; policemen were shot, and gun shops were emptied by crowds of men. Armed bands of desperadoes paraded in front of the residence of the commanding general of U.S. troops; another group thought to be associated with the Knights of the White Camellia drilled at night. A mob threatened the superintendent of police, and boards of registration were attacked. There were a number of reports of extreme acts of random violence and murder.[18]

In November 1868, Mumford McCoy, a freedman and resident of St. Helena Parish, testified that he was confronted by a group intending to prevent him from continuing to influence other freedmen. "We are the white men of this country," he was told, "and we intend to rule it." The men told him that "the Yankees and the carpetbaggers can do you no good." When asked what his assessment of the feeling of the white people toward the colored people, McCoy answered, "they say and they are determined to kill every radical, white, or black."[19]

At St. Landry Parish in November, former Union soldier S. A. Miller, a Republican poll worker and active in promoting demonstrations of loyalty to the United States and support for the Reconstruction government, provided his view of the hostility of white southerners: "I am satisfied by the appearance and determination of the people that a man could not openly declare himself a Republican."[20] William Dunn of St. Martin's Parish testified that "all kinds of intimidation were used by the democrats toward the colored people to induce them not to register; but if they did, they must vote the Democratic ticket, or they would be driven off the plantations. The colored people are frightened to death."[21] In October, in Saint Bernard Parish, an act of arson sparked retaliation as "bodies of armed white men notwithstanding the presence of United States troops who were there to aid the civil authorities, prowled the parish killing about 20 and maltreating every freedman they could find" and robbing residences.[22]

Between 1869 and 1872, the U.S. military presence declined, leaving one regiment of infantry to secure the state. Outside of outbreaks of violence during elections, the state was generally quiet. This all changed when Warmoth ran afoul of a clique of federal appointees in New Orleans, among them William P. Kellogg, a Union officer who served a short time in the war, and through a personal connection to President Lincoln, became the customs collector at the port of New Orleans. After Louisiana's readmission, he was appointed a U.S. senator. In 1872, Warmoth attempted to leverage his connections with Democrats to defeat Kellogg's bid for governor, supporting Kellogg's opponent, John McEnery. Although it appeared that McEnery had defeated Kellogg, accusations of fraud led to a dual inauguration in January of 1873, even though President Grant had recognized Kellogg as the new governor. Maj. Gen. William H. Emory, commander of the Department of the Gulf, had stationed troops in New Orleans to deter the factions from engaging in open warfare. Under Grant's orders, Emory allowed the two factions to coexist.[23]

This unsettled political condition opened Phase III of asymmetric warfare. In February of 1873, the shadow governor McEnery appointed Fred N. Ogden, a former Confederate officer, as commanding general of the militia. Ogden rapidly organized an armed force of 500 men in New Orleans and developed plans to eliminate the Metropolitan Police, a mixture of blacks and whites who were now part of Longstreet's state militia. An open battle in the city's streets erupted in March that was halted by the intercession of federal troops. Shortly thereafter, the state forces surrounded the building where the McEnery legislature was meeting and forced them to vacate, leaving Kellogg as the sole head of government.[24]

Kellogg attempted to solidify his control of the state by offering appointments to supporters for key parish administrative positions. In Grant Parish, he promised positions to the Republican faction, but McEnery had already appointed a judge and sheriff. The Republican faction occupied the courthouse in Colfax and prepared to defend it. Both factions armed themselves, and rumors of depredations spread among both whites and blacks. Blacks fled from the surrounding countryside for the apparent security of Colfax. The town was patrolled by armed blacks, a number of them former members of a local defunct state militia company. Several skirmishes occurred as white detachments tested the defenses, and both sides sought assistance from surrounding parishes. Many former Confederates and members of the Knights of the White Camellia responded, and a small artillery piece had been procured. About 140 men had committed themselves to action, recognizing that it was an all-or-nothing effort. The blacks had dug a modest entrenchment and had committed themselves to a defense, even as the white Republicans deserted them. A well-executed tactical assault rapidly overwhelmed the defenders. Those who attempted to escape were chased down and killed, and those who surrendered were also later deliberately killed. A total of 59 bodies were discovered in the aftermath as U.S. troops took control of the area. As a demonstration of will, the Colfax massacre had a powerful effect throughout the lower South. It was the harbinger of what a race war would entail. Organized groups of armed blacks, whether militia or otherwise, would be challenged with military force, and no quarter would be given.[25] The *Shreveport Times* captured the impetus of the mobilized white Southern identity behind this particular manifestation of asymmetric warfare:

> Everywhere the white people have cast off the apathy and despair which have for years crushed them into the dust, and enter upon a movement with a resolution and an enthusiasm that must accomplish the most gratifying results. Unless the insolent encroachment on the rights, political and social, of the white people of Louisiana is ended, the day of the *'irrepressible conflict'* will *come when physical force shall solve the political problems in Louisiana politics.*[26]

Between May and August of 1874, a loose collection of paramilitary units known as the White League grew in strength and power. What made these organizations different from the Knights of the White Camellia was their military discipline and leadership. The White League identified themselves as oppressed taxpayers arising from popular discontent who wanted to restore honest and intelligent government to Louisiana. Men united to join the White League to defend their property and families from Republican threats in the form of the Metropolitan Police.[27] The creed of the White League, according to a St. Landry Parish newspaper, was "the union of all white men in opposition to Radicalism and negro domination." The White League in Avoyelles Parish was dedicated to "setting forth grievances and with firm determination to relieve themselves of the burdens that are grinding out their very existence."[28]

The Caddo Parish taxpayer's association held a meeting in July of 1874 and adopted a series of resolutions. "We attribute all the trouble here threatening us to the teachings of the *white carpet-bagger*," stated one of the resolutions printed in the *Shreveport Times*, "and we *intend to strike this evil at the root if we are forced to desperate measures*."[29] Also in July, at Greensborough, in Lincoln Parish, a state senator sent a message to the U.S. marshal in New Orleans requesting half a company of U.S. troops or Metropolitan Police to be sent to the parish to preserve order. He reported that heavily armed men of the White League ordered a Republican club to disperse in 10 minutes or it would be attacked. The senator was told that if blacks paraded in Lincoln Parish, displayed the national flag, beat a drum, or organized in any sizable number, the senator would be killed and the blacks fired on.[30] In October, the White League drove out the most prominent Republicans in Bienville; in Bellevue, a White League force of 200 ordered the parish judge out of town; and in Iberia Parish, the White League demanded the resignations of parish officers.[31]

A detachment of the U.S. 7th Cavalry was stationed at Shreveport in the Upper Red River District beginning in October 1874 and remained there for 15 months; however, it was restricted from interfering with local affairs. Maj. Lewis Merrill, who had played such a prominent role in South Carolina during the Ku-Klux counteroffensive, was the commanding officer. After a month of duty, he reported in November that the large majority of whites were bent on eliminating both black and white Radical leaders, whether by assassination or by driving them out. Election results mattered little in this effort. "I do not see that the military can do anything," he concluded, "except furnish the strength to arrest."[32]

The initiative had passed to the White League, opening another opportunity for a direct attack on the Republican government in New Orleans. There, Ogden had rebuilt his force into a well-disciplined organization, led by Confederate veterans and with a large number of veterans in the ranks.

He called his unit the Crescent City White League, and the purpose of this unit was to overthrow the Kellogg government if the opportunity arose. Metropolitan Police reports estimated the White League in New Orleans at 4,150 men. Company size units (including four companies of artillery) of 100–300 men drilled without arms in the street, in vacant lots, or at their club rooms in the wards of the city.[33] Alarmed by the growing force in their midst and alerted that shipments of modern arms were being delivered to the White League, the Metropolitan Police began to search cargo deliveries at the city wharves. Needing the weapons for any assault on the government, Ogden conferred with Davidson B. Penn, (who was McEnery's lieutenant governor and could lay claim to legitimacy), and a decision was made to seize the opportunity offered. The majority of federal troops in the city had been moved to Mississippi, which prevented any significant intervention. A mass demonstration would call for the resignation of Kellogg, followed by an armed assault intended to capture the statehouse and install Penn as the acting governor.

The execution of the operation was conducted with exceptional military efficiency on September 14, 1874. To isolate the city from outside support, the telegraph lines were cut, and key city government offices were seized. Key streets were blockaded to isolate the state and federal government buildings from the rest of the city. The mass demonstration was held at a point where the crowd of about 5,000 could easily gather and just as easily disperse. A detachment moved out to capture a weapons shipment under control of the Metropolitans. Kellogg was informed that he was being deposed and ordered to surrender. Kellogg refused and fled to the safety of the U.S. Custom House with its small detachment of federal troops.[34]

James Longstreet as the commander of the state militia had about 1,000 well-armed men at his disposal, as well as Gatling guns and two 12-pound artillery pieces. He employed his black militia units around the statehouse and sent a 500-man Metropolitan force with mounted cavalry, the Gatling guns, and artillery to strike directly at the crowds and sweep north to clear barricades and regain control of the city. Longstreet's troops found themselves outflanked to the south, and a detachment with artillery and a Gatling gun holding a hasty blocking position by the levee was overwhelmed in a disciplined charge punctuated with the rebel yell. The militia broke and scattered as the White League, now fully armed with newly distributed weapons and captured equipment, controlled the city, but several heavily defended strongpoints still held out. Overnight, the black militia melted away, leaving Kellogg and Longstreet isolated in the U.S. Custom House.[35]

The following day, President Grant issued a proclamation ordering the armed groups to disperse and submit to Governor Kellogg and the legally constituted authorities of the state. Federal units rushed to New Orleans wisely took stock of the men who were better armed than they were and decided to leave matters as they were. After a victory parade, the White League units awaited orders. General Emory arrived in the city and ordered McEnery to return city property to the authorities. The buildings were turned over, but the cannons and captured weapons remained under White League control. Emory ordered the city be placed under military control until he could return Kellogg to the statehouse on September 18. By the end of the month, 800 U.S. troops occupied the city. Blood had been shed— Longstreet's Metropolitans ended up with 11 men killed and 60 wounded, and the White League suffered 21 men killed and 19 wounded.[36]

In an interview with a Northern reporter four months later, Ogden claimed that the White League in New Orleans was organized in self-defense against the black militia and included "almost every white man not a Republican in the city." He described them as "veterans who have been under fire." In a reflection of the Southern collective identity, Ogden identified the enemy. Kellogg and nearly all his officers were carpetbaggers who held office illegally through fraud and were protected by federal troops. Ogden defended his actions, stating simply, "I was acting in concert with the people."[37]

As asymmetric warfare, Ogden's action was highly symbolic, directed at the heart of the Congressional strategy of Reconstruction. There was no need to seize and hold power. It would have been too difficult and would have led to open battle with federal troops. By demonstrating a capability that could be employed at will against the Kellogg government, Ogden, Penn, and McEnery had gained the psychological advantage, and by quietly obeying General Emory's orders, the White League also demonstrated that only federal power could sustain this unpopular government imposed on them by the Congress.

This demonstration of asymmetric warfare revealed its influence on Northern public opinion. D. M. A. Jewett, the U.S. commissioner for the District of Louisiana, Lincoln Parish, sent a letter to the *Boston Daily Globe* just a few days after Kellogg had been reinstated as governor. "Until the heavy hand of the nation is felt here past and present crimes punished, and treason and rebellion against State and nation 'made odious' and perilous," Jewett fumed, "the 'state of affairs in Louisiana' will remain what it is." The *Daily Globe* printed the letter and responded with an editorial that sought reconciliation. A "policy of proscription and punishment" was not useful. "We want the leaders in the rebellion and those who once sympathized with treason to come back and take a part in the administration of affairs. It is only with their help that the breaches of war can be healed and the fruits of peace fostered." The *New York Tribune* responded to the events in Louisiana, commenting that the use of military force was intended to support the Republican political agenda and to keep Republicans in power.[38]

Adelbert Ames was a West Point graduate from Maine and, who began the war as a captain and fought in most of the major engagements of the

Army of the Potomac, ending the war as a major general at the age of 29. Ames was the commander of the postwar federal occupation force in Arkansas and Mississippi designated as the Fourth Military District. President Johnson gave him the additional duty as the provisional governor of Mississippi. Ames married the daughter of former Union general Benjamin F. Butler and now a prominent Radical Republican politician in the House of Representatives. Butler was one of the most hated men in the Confederacy. Because of his important political connections, President Lincoln promoted Butler to the rank of major general. As the military governor of occupied New Orleans in 1862, Butler earned undying enmity for his high-handed treatment of the population and his reportedly shady activities. During the war, he commanded at a level far beyond his capabilities. He was the author of a number of battlefield failures in the latter stages of the war that sorely tried the patience of Gen. Grant.

Like his father-in-law, Ames embraced the Radical agenda of social and political equality for the freedmen and became a leader in the new Republican Party in Mississippi. After Mississippi was readmitted to the Union in 1870, Ames served as the state's senator in Washington. Ames ran for governor and was inaugurated in 1874. He faced an angry white population that was opposed to the high taxes that had been imposed by the previous Republican administration. Ames promised to lower taxes and end corruption, and he downplayed rumors of an impending race war, declaring it both unlikely and abominable. The economic hard times that hit Mississippi in 1874 ruined any goodwill that Ames had tried to build. Suffering was widespread, and the tax burden only added to the misery, leading to the confiscation of one-fifth of the most productive land in the state. Contrary to his promises, taxes rose and state expenditures were never curtailed. Governor Ames surrounded himself with a small coterie of advisors, all of whom were considered carpetbaggers, and isolated himself from other Republicans as well as Democrats.[39]

The lack of action led to the creation of taxpayer organizations focused on opposing the corrupt Republican administrators, most of whom were blacks who dominated the key local government positions in the state. In Vicksburg, the largest city in the state, economic conditions reached a crisis point at the time of local elections in August. White Southerners and a number of transplanted Northerners, many of whom were Union army veterans, allied to challenge Republican dominance. Men began arming themselves and demonstrating. Ames appealed to President Grant for federal troops to prevent violence, but with the 1874 congressional elections not far off, Grant demurred. Whites carried the municipal election by 350 votes, the first time the city had not been under black control since the Civil War. "No aggressive violence was used," a Louisiana newspaper noted, "but very definite intentions were given that such a resort would be met with a vigor, a courage and resolution."[40]

Between August and December of 1874, the atmosphere in Vicksburg became increasingly tense. Whites controlled the city government, but blacks controlled the Warren County government and continued to support corrupt local officials. A taxpayer's association led by John D. Bland, a former Union army officer, harnessed popular discontent over white taxpayers being ruled by black nontaxpayers. The association complained that radicals controlled every department of government and used the courts as a means of oppression. The call was for white unity—there could be no middle ground against the Radicals.[41]

The immediate target of taxpayer wrath was Peter Crosby, the black county tax collector and sheriff. Crosby had already created public fury by blatantly withholding evidence against three black officials indicted for fraud. Crosby also resisted making a new bond to secure tax funds collected after his original bond was found to be worthless. On December 2, an angry crowd assembled and confronted Crosby, calling on him to resign or face an uncertain fate. Crosby resigned, left the city, and sought assistance at the governor's mansion. Ames believed Crosby faced only a "riotous and disorderly" assembly and was neither willing to use the state militia nor did he believe it necessary to appeal to the president for federal forces. Instead, Ames sent Crosby back with two militia officers with instructions to organize a posse and retake his office.[42]

Crosby distributed a handbill that was read in the county's churches on December 6. He claimed he resigned under compulsion, "the result of base coercion on the part of an armed mob of the most bitter and relentless of our enemies." The next day, Crosby dutifully organized a collection of about 125 blacks and began an advance on the city.[43] The mayor of Vicksburg declared martial law and prepared to defend the city. Whites from the surrounding areas assembled, along with about 160 men from the Louisiana White League. As Crosby's group approached the city, it became clear that there was no hope for the sheriff to return to his post. After a short standoff, the blacks agreed to withdraw but were fired upon. This led to a general attack that became a widespread assault on Warren County blacks in general, who were believed to have supported Crosby's march on Vicksburg. More than 300 blacks and 2 whites were killed.

The Mississippi State Senate, spurred by the governor, sent a message to President Grant on December 19, which drew a stark picture of anarchy and revolution. Large numbers of armed men from adjacent states had invaded Mississippi to provide support to "lawless and riotous persons" in Vicksburg. The county courts were "paralyzed," and there was no force available that was capable of gaining control of the situation. The president was asked to employ military force to suppress the violence in order to "guarantee to all citizens the equal and impartial enjoyment of their constitutional and legal rights." On January 5, General Sheridan sent a message to Governor Ames indicating that a company of troops would be dispatched to Vicksburg. About 50 soldiers arrived in the city, and Crosby was reinstated.[44]

Like New Orleans, the asymmetric advantage of will was clearly demonstrated in Vicksburg. Winning popular support to justify the use of force to reorder the political landscape as a legitimate expression of taxpayer frustration with corruption and incompetence was far more important than success or failure. A Jackson, Mississippi, newspaper shortly after the arrival of federal forces in Vicksburg noted with satisfaction that "the negroes remembering the lessons of Crosby's foolishness and Ames' treachery . . . will not repeat their experiment, and the white leagues are certain not to molest them."[45] The 300 dead blacks scattered in Warren County demonstrated that mobilized white Southerners were willing to initiate a race war when challenged. Ames was completely unwilling to take any action that would lead to such a horrific outcome. Instead, his feeble response was to send Crosby and his posse to certain destruction. This incident, coming on the heels of the New Orleans street battles, contributed to the Southern dominance of the information battlefield. Indeed, the action of the Ames government to bring in federal troops to restore a Republican government official was widely condemned. One newspaper accused Ames of destroying "thrift, decency, prosperity, feeling, religion, morality, and . . . good government."[46]

Vicksburg was a catalyst for further action representing Phase III of asymmetric warfare to overcome Radical control of the state. Albert T. Morgan, the Yazoo County sheriff, was a former Union soldier and dedicated Radical Republican who came to Mississippi to find his fortune. Although his attempts were unsuccessful, he had political influence as the Republican state convention chairman and strong black support. Winning the election for sheriff in 1873, he had to take his office from the incumbent with the threat of armed violence that led to the death of an opponent. Governor Ames arranged for a friendly judge to hold the trial, and Morgan was found innocent and returned to his duties. Morgan collected taxes and began using the money to fund schools for his black supporters (and to employ his Northern mixed-race wife, who was a schoolteacher). The economic hard times that continued for the next two years only increased white resentment as their taxes were an increasing burden, and all the benefits appeared to be going to the blacks.[47]

In September 1875, as the election approached, Morgan organized and armed a personal bodyguard and began drilling them at night. Predictably, whites formed their own armed units, and a confrontation was inevitable. A short fight broke out at Satartia, Mississippi, which led to about 800 heavily armed men assembling and forcing Morgan to flee for his life. Governor Ames issued an ineffectual proclamation ordering armed organizations to disband and submit to lawful authorities, but by then, Yazoo was under control of armed men defying the state government and preparing to defend the town.[48]

Shortly thereafter, about 2,000 people, mostly black, were in attendance at a Republican political rally in Clinton, Mississippi. A small group of white men were present and heard the speech of the Democratic candidate for state senate. When the Republican candidate began speaking, he was harassed, which led to a confrontation with men brandishing pistols. At about the same time, a black marshal attempted to enforce the restriction against alcohol on the grounds, which also led to a confrontation. In the space of a few minutes, these separate confrontations led to blacks and whites engaging in a pistol battle. Four blacks were killed in the approximately 15-minute exchange of fire. Two white men were also killed, and their bodies badly mutilated; another white man, a bystander, was also killed and mutilated. The white survivors arrived at Clinton, and, almost immediately, an armed force was organized and moved out to sweep the grounds where the incident took place. A call for assistance brought nearly 500 men to Clinton, who set up a hasty defense and sent out patrols. Two blacks, believed to be scouts, were killed. Reinforcements arrived by train from Vicksburg and moved out into the countryside, killing any blacks they encountered. Anywhere from 12 to 50 blacks were reported killed. A ceasefire was brokered and peace returned.[49]

On the heels of the news in Yazoo and Clinton, Governor Ames sent a telegram to President Grant, reporting that "unauthorized and illegal armed bodies" in Yazoo and Hinds counties were conducting attacks and that the state authorities were powerless to stop them. Claiming the necessity for immediate action to protect lives, he requested federal troops be sent to the state. Grant was reluctant to dispatch troops. The political climate in the North was growing hostile to the idea of repeated interventions in the Southern states. The soldiers would be seen as nothing more than a political symbol intended to whip up Republican voter ire for the upcoming election.[50]

When his request was denied, Ames decided to employ the black militia. The reaction was immediately met with a call to arms. A Jackson, Mississippi, newspaper announced the impending arrival of the black militia and proclaimed, "the scent of blood is in the air." Governor Ames was willing, according to the newspaper, to unleash "civil war, rapine, a war of the races." The appeal went out for "every citizen to hold himself in readiness" to join armed units that were organizing to resist. Gen. Christopher C. Augur, commander of the Department of the Gulf, advised Ames against this action, fearing that it would open a race war that would spread throughout the Gulf States. Ames prepared to use two companies of black militia to restore order and return Morgan to his position, but when Morgan refused to return, the operation was called off.[51]

Ames sent reports to President Grant of unauthorized and illegal armed bands employing violence throughout the state, but Grant would not budge, unwilling to pay the political cost of dispatching troops to Mississippi again. Without federal forces, Ames faced the prospect of employing the state's all black militia force against a population that demonstrated the capability to organize and assemble large groups of armed men, and who appeared to be looking for an excuse for an all-out war. This explosive situation was settled through what was essentially a cease-fire agreement in October, brokered by a representative of the U.S. attorney general. Black militia units would be disbanded, and two other units would be allowed to be held in reserve. In return, white armed groups would refrain from further violence.[52]

In the 1875 elections, the Democratic Party in Mississippi won four of six congressional seats, secured a large majority in the state legislature, and regained control of most county governments. Out of 71 counties, 22 had a major loss of Republican voting strengths, largely due to the influence of armed groups. When the Democratic legislature convened in January 1876, impeachment charges were brought against Governor Ames and Lt. Gov. Alexander K. Davis. Davis was impeached and removed from office. In return for the legislature dropping impeachment charges against him, Ames agreed to resign. Ironically, the impeachment charges facing Ames were for "abuse of power and violation of the law directing Peter Crosby to return to Vicksburg and take possession of the office of Sheriff by force of arms," and fostering a "conflict" and "war of the races."[53]

In Louisiana, the outcome of the state elections resulted in an even split between Democrats and Republicans, with five seats undetermined to be resolved by the legislature when it met in January 1875. The situation was serious. William W. Belknap, the secretary of war, acting through the president, tasked Lt. Gen. Philip H. Sheridan, commander of the Division of the Missouri, to ascertain the situation in Louisiana and Mississippi and to pay special attention to the situation in New Orleans. He was given wide latitude from the president "to ascertain the true condition of affairs, and to receive such suggestions from you as you may deem advisable and judicious."[54] He was also authorized to take command, if necessary.

Sheridan had few friends in Louisiana. He had been the commander of the Fifth Military District in 1867 and had carried out the requirements of the Congressional Reconstruction Acts with military efficiency, involving himself in state politics by replacing state officials who he believed were not cooperating with military authorities, including the governor. President Johnson relieved Sheridan of command of the district to avoid antagonizing white Southerners any further. In fact, Sheridan himself admitted in 1867 that "nearly every civil officer within my command was either openly or secretly opposed to the law and to myself." A Pennsylvania newspaper made a similar observation, describing Sheridan's name as "a red rag to nearly the whole population of the Gulf States."[55]

Sheridan arrived a few days before the legislature convened. There were signs that the White League would attempt to intervene. A few days before the opening of the legislature, R. H. Marr, one of the leaders of the September 1874 insurrection, declared that, "we are willing to lose our lives in defense of our rights, willing to shed our blood if we can hope thereby to secure freedom and justice for our posterity."[56] To prevent a repeat of the September 1874 street battle, the statehouse was ringed by Metropolitan Police and federal troops. Between 12:00 noon, when the legislative session was officially opened, and 2:00 p.m., the Democrats had staged a fantastic array of legal legerdemain that not only took control of the session but also certified Democrats to fill the five vacant seats, and they had federal troops clear the House chamber. The Republican legislators, completely outflanked, appealed to Governor Kellogg, who made a formal request for federal troops to clear the statehouse of the illegally recognized Democrats. Again federal troops entered the House chamber with fixed bayonets, this time coming after the Democrats. As the five members were escorted out, they were followed by the Democratic legislators, protesting the use of armed force. The Republicans immediately took control, assuring a Republican majority.[57]

At this point, Sheridan stepped into the breach. He perfunctorily assumed command of the Department of the Gulf from General Emory, who had been in command since 1871, and sent a series of bombastic dispatches to the secretary of war, reporting on the situation, while overstepping his professional boundaries as a military commander by making policy recommendations to the Congress and the president. On January 4, Sheridan sent his first dispatch to the secretary of war, describing the situation. In his assessment, "a spirit of defiance to all lawful authority" existed, and "the civil government appears powerless to punish or even arrest." The day after the fracas at the statehouse, Sheridan again reported to the secretary of war:

> I think the terrorism now existing in Louisiana, Mississippi, and Arkansas could be entirely removed . . . by the arrest and trial of the ringleaders of the armed White Leagues. If Congress will pass a bill declaring them banditti, they could be tried by a military commission. It is possible that if the President would issue a proclamation declaring them bandits, no further action need be taken except that which would devolve upon me.[58]

A few days later in another dispatch to Belknap, Sheridan complained that those who murdered political targets in Louisiana were "regarded as heroes" in their localities and "by the White League and their supporters."[59]

These ill-considered messages largely doomed the Congressional Reconstruction strategy. As one of the most prominent heroes of the Union army in the Civil War, Sheridan raised his profile to the national level by taking the highly visible position of command and associating himself with federal troops with bayonetted rifles clearing the statehouse of Louisiana. He compounded his error by having his dispatches sent to the local newspapers. In doing so, Sheridan became the symbol and focus of widespread outrage against what many in the country believed Reconstruction had become. The *Daily Argus* of Rock Island, Illinois, announced that Sheridan's actions initiated a new Republican threat to all citizens. "The military arm is outstretched to-day to seize the power of the republic and to cut a bloody pathway to a new form of personal government."[60] The *Daily Dispatch* in Richmond, Virginia, took great delight in quoting the Philadelphia *Telegraph* referring to Sheridan's dispatches: "When a soldier drops the sword and takes up the pen he is very apt to make a fool of himself." The Richmond newspaper also observed, "In no light that we can view the course of the Administration toward Louisiana and the South can we consider it anything but weak and tottering. The army alone gave it supremacy over the States, and *nothing but the army can maintain its existence after the present term*."[61]

The *New York Herald* announced on January 9 that a mass indignation meeting would be held in the Great Hall at Cooper Union. The *New York Times* reacted to Sheridan's dispatches with a declaration that "nothing like it has ever been seen before under any Constitutional Government." Sheridan, the *Times* declared, "could scarcely have shown greater ignorance or disregard of law," and the newspaper denied Sheridan's actions had any connection to the Republican Party.[62] A Pennsylvania newspaper compared Sheridan to Cromwell as a "military usurper." A *Chicago Times* correspondent in New Orleans reported that Sheridan had "awakened a whirlwind of indignation, which can be heard at every street corner, in violent tones."[63]

Ten days after Sheridan's first dispatch, a reporter for the *New York Tribune* interviewed a number of members of Congress regarding Sheridan's actions. Most senators who were approached had nothing to say. However, Glenni W. Scofield, a Republican member of the House from Pennsylvania, made remarks that indicated an appreciation for the asymmetries of will and organization that existed in Louisiana among white Southerners. He stated that the opposition to the Republican government had all the intelligence, the energy, and the ability to carry out their goals. In addition, he believed the opposition had the advantage in wealth, courage, and experience in public affairs. The Republicans, he thought, were no match for them. There would be no peace until the opposition took power in the state.[64]

A few days earlier, Democratic representative Samuel Cox of New York stated that peace could only be secured by the withdrawal of federal troops. Democratic representative Stevenson Archer of Maryland blamed the corruption, fraud, and misrule of the Republican government in Louisiana as the source of the unrest. The archbishop of New Orleans and the bishop of

Louisiana published an open letter to the American people claiming that "these charges are unmerited, unfounded and erroneous, and can have no other effect than that of serving the interests of corrupt politicians, who are at this moment making the most extreme efforts to perpetuate their power over the State of Louisiana."[65]

It was, however, Sheridan's description of the White League as "banditti" that raised the greatest public ire. A Maryland newspaper derided Sheridan for "pretending to understand the feelings and state of mind of the people of New Orleans after a stay among them of three days," as well as "stigmatizing the people of Louisiana as a banditti."[66] A New Orleans newspaper reacted with a strong expression of collective identity and defiance:

> We came from sires who have shed luster upon every page of worthy history of this broad land; we love our civilization; we cling to our traditions and the habits and customs of our homes and hearthstones; we prize liberty [and] we are willing to lay down life itself to secure it for ourselves and our children and for all this, we are denounced as *bandits* and *murderers*![67]

Sen. John Brown Gordon of Georgia, one of the most famous commanders in Lee's Army during the Civil War, most clearly voiced the essential element of the Southern-Confederate collective identity that would serve as the dominant and decisive theme of the information battlefield:

> Men came down among us who have no interest in common with us; they hold the offices, make our laws, levy our taxes, spend them, and whenever we endeavor lawfully to recover our rights, we are stigmatized as assassins, as murders, as semi-barbarians and as disloyal to the Federal Government. When is this misjudging of the Southern people to stop?[68]

A few days after Gordon's speech, President Grant sent a message to the Senate asking for some clarification on his ability to act under the Fifteenth Amendment to address the situation in Louisiana. An investigating committee was sent to Louisiana chaired by William A. Wheeler of New York. In a telling indication of the inability of the dominant actor to enforce its will, the Congressional delegation began to negotiate for a political settlement. The Democrats were given control of the lower House, while the Republicans controlled the Senate. In return, the Democrats agreed not to threaten Governor Kellogg's position. This was a key turning point in Phase III of asymmetric warfare. For the first time, the dominant actor began to define acceptable terms for establishing a political equilibrium. The weaker actor was now capable of seeking additional compromises and accommodations throughout the South.[69]

The Congress made one last symbolic attempt to pursue the strategic goals of Reconstruction. In March of 1875, President Grant signed the Civil Rights Act providing to citizens of every race and color equal access to public accommodations, facilities, and transportation. It had been the personal crusade of Sen. Charles Sumner, who intended it to be the crowning achievement of the Congressional Reconstruction strategy. He had battled in favor of the measure for nearly five years before a reluctant and divided Republican Congress finally passed a watered-down version a year after Sumner died in March 1874. A clear indicator of the declining will of the Congress, the law was largely a symbolic act in tribute to Sumner. Although it initially created a great deal of consternation and concern in the South, the law was widely ignored in practice and lasted only nine years before it was struck down by the Supreme Court.[70]

Meanwhile, white Southerners demonstrated a growing lack of fear of the national government as state Republican power continued to dissolve under the shadow of federal military presence. A Republican state legislator in Austin, Texas, was met by an armed group that included the sheriff and his deputy, demanding that he leave the county or be killed. The only reason given was that he was a "damned radical." In Louisiana's St. Martin Parish, the White League forced the Republican sheriff to resign; the U.S. marshal was unable to make any arrests without having the support of federal troops. Blacks were discharged from jobs for voting Republican. White Republicans were publicly and socially ostracized. "Our opponents are thoroughly organized," a pro-Republican newspaper reported, "unless [black voters] are previously excited and drilled, one-half of them will not come to the polls, and a large per cent of the remainder will vote the white man's ticket."[71]

The shock of violent armed resistance in the South caused Northerners to reexamine the direction and goals of the Congressional Reconstruction strategy. In reviewing the events of the past, the *New York Herald* pointedly concluded that the Northern spirit of revenge had led to the economic ruin of the entire region, resulting in its inability to establish credit and pay debts. The *Herald* claimed that the burdensome financial indemnity Germany imposed on France in the aftermath of the Franco-Prussian War in 1871 was minor compared to the punishment inflicted on the South after the Civil War. The entire strategy of Congressional Reconstruction was flawed. Echoing the *New York Times*, the *Herald* claimed that by failing to include "the real sons of the South," the success of Reconstruction had depended on "ignorant freedmen" and "unscrupulous white adventurers," which only resulted in corruption, unrest, and disorder. If peace and harmony were ever to return to the Union, the real sons of the South must govern the South, "*otherwise there will be no peace.*"[72]

The weaker actor had achieved decisive success over the dominant actor. The Congressional Reconstruction strategy was being discredited, the costs of sustaining it were perceived as too high in terms of the nation's financial health, and the folly of marginalizing white Southerners had become all too clear. Pressure was mounting to find some practical means to transition power from the corrupt, ineffective, and ill-favored Reconstruction governments to governments that could establish peace and stability with some degree of protection for all citizens. The Grant administration, mired in scandal, was attempting to keep its hold on power and needed the support of the Republican-controlled states in the South. For the near term, there would be no change in policy or strategic direction of Reconstruction.

White mobilization in the states throughout the South between 1870 and 1875 was based on personal honor and the protection of home and family, which were the nonnegotiable factors of Southern-Confederate collective identity. Prior to the war, white Southerners had maintained a bedrock belief in the principle of local self-government and the principle of the concurrent voice that protected the minority against the supremacy of the majority, relying on the power of the Constitution to protect their interests within the federal union. The triumph of the Union through the total defeat of the Confederacy had eliminated all of these beliefs. The Constitution had been changed fundamentally, and black men were no longer subordinates, but citizens with political rights. They had been elevated to essential political actors in order to sustain an alien government imposed on the vanquished by the victor. In this new political environment, Southern whites began to reassert their unshaken beliefs in personal freedom, independence, and self-rule that represented the essential rights and privileges of white citizenship. As the Southern-Confederate collective identity was mobilized, it was clear that these rights and privileges had to be defended to whatever extreme was necessary. The willingness of Southerners to unleash such savagery on blacks especially was a clear signal to all that the rights and privileges of citizenship in the new order were only to be guaranteed by violence. The dominant actor's constitutional amendments, acts of Congress, presidential proclamations, and state laws declaring the social and political rights of persons of all races and colors meant nothing; blacks and their white allies would have to fight for them with the same level of dedication and ferocity as the weaker actor was willing to fight for those same rights.

The anger, resentment, and fear that usually culminated in violent reprisals whenever blacks assembled or displayed hostile intent were the inherent factors that drove Phases II and III of asymmetric warfare. These asymmetries of will and violent action could not be matched, nor could the weaker actor's dominance of the information battlefield, which was a critical factor that drained the will of the dominant actor. As the mobilization threshold of the Southern-Confederate collective identity continually grew stronger during this period, the ability of blacks and Republican leaders to defend their political gains grew weaker. The level of resistance could not be curbed or reduced as long as blacks continued to play the role of an essential political actor and as long as Republican governments and their federal bayonets remained.[73]

In Phase III, white Southerners took an indirect military approach to dislocate a number of Republican state governments by negating their enforcement power and a direct parallel political approach to challenge their legitimacy. Southerners came to understand that time was the political vulnerability of the Congressional Reconstruction strategy. By drawing out conflict through episodic violence and maintaining an unsettled political security situation in the states, neither the federal government nor the Republican governments had an effective response to the mobilized collective identity, which demonstrated a high level of commitment to resistance and a willingness to escalate confrontations to an open race war. The longer the situation continued, the higher the political, psychological, and economic costs rose for the dominant actor and the more difficult it became to sustain the political will to continue the effort to achieve the desired strategic goals. The upcoming election of 1876 would serve as the final blow to the dominant actor's strategy.

8

Asymmetric Warfare Phases III and IV: Bulldozers, Red Shirts, and Equilibrium (1876–1877)

> See the glorious banner waving,
> Proudly in the air;
> Victory will crown the efforts
> Of those who do and dare.
> Song of the "Hampton Boys," 1876[1]

By 1876, the Congressional Reconstruction strategy had been largely defeated. The power of the weaker actor had become stronger than the dominant actor because the interest of the weaker actor in achieving its goals remained stronger relative to the dominant actor's interest in achieving its strategic goals. Tennessee, Alabama, North Carolina, Georgia, Arkansas, and, most recently, Mississippi, had returned to the control of white Southerners by various forms of asymmetric warfare, employing and encouraging a level of violence that prevented the Republican state governments from establishing legitimacy. The Southern dominance of the information battlefield had crippled the Northern will to sustain the strategic goals of Congress.

The Northern public's increasing frustration with the Grant administration's "bayonet rule" approach to the states under Republican control greatly limited the options of the president and congress. By 1876, the Republicans were widely perceived as initiating their policies out of hatred of the Southern people, while taking every opportunity to oppress and humiliate the South. The Reconstruction governments had been imposed by bayonets against the will of the people, which encouraged misrule by the dishonest men placed in power. The result was, as one Northern newspaper maintained, "partisan strife and political bitterness."[2]

After 11 years of unchallenged political power, the Republicans had little to show for their efforts in the South. Anticipating the upcoming presidential contest, the *Los Angeles Daily Herald* expressed the opinion that another Republican victory would only make the current situation in the South far worse: "We can but think that the efforts of the Radical extremists to re-open dead issues will prove so nauseating that they will end in emphatic failure."[3] Despite years of effort, complained an Ohio newspaper, the Republicans still claimed that the South was ungovernable and that the murder of blacks was rampant. Even if these claims were true, the editor noted, the evidence appeared to indicate that the Republicans were, by their own admission, unfit to hold power. A North Carolina newspaper had a more blunt assessment: "The longer the decaying carcass of Southern Republicanism lasts, there are fair-minded people everywhere impressed with its weakness and worthlessness."[4]

The Republican governments of Louisiana and South Carolina maintained a tenuous hold on power, anchored by black votes and the nominal, but largely symbolic, presence of federal forces. Florida, although still under Republican control, was already largely moving toward a balance of power based on mutual economic interests between Republicans and Democrats. In Louisiana and South Carolina, whites who had formed political alliances with freedmen years earlier had, for the most part, drifted back to the Democrats. Yet in these states, the unified white vote by itself was not sufficient to unseat the Republican governments. As the watershed state and national elections of 1876 approached, white Southerners in Louisiana and South Carolina adopted a new asymmetric approach. This approach emphasized conventional political activity, but it was backed by a paramilitary bulldozer element, which operated with complete impunity to neutralize the state militia and degrade the voting strength of blacks to ensure the triumph of the Democrats.

The bulldozer element emerged in the summer of 1876, in anticipation of the elections. Unlike the Ku-Klux phenomenon, the bulldozers had a clear political purpose and direction. The term "bulldozer" originated with blacks in Louisiana to describe the usually mounted and heavily armed men who employed strong-arm tactics as well as violence to establish control over the voting population and Republican officeholders.[5] In Louisiana, the White League became the basis for a highly organized force in being, operating with a high level of efficiency and unity to set the conditions for success by preventing Republicans from voting or preventing individuals from encouraging or organizing others to vote Republican. After reviewing a number of Louisiana newspapers, an Ohio newspaper provided a report to its readers that described the bulldozer element in the state as armed units that operated in the open, were directed by some type of central military command, and represented a constant menace to Republicans. The purpose of the bulldozers was to employ whatever means necessary to invalidate the power of the Congress to enforce its will.[6]

In South Carolina, the bulldozers comprised existing paramilitary rifle and saber clubs that had been formed throughout the state in the aftermath of the federal counter Ku-Klux sweeps in 1871. For all intents and purposes, these groups displayed the military organization and discipline of cavalry units as they conducted patrols in various parts of the state, usually with the purpose of confiscating arms from blacks. Adopting a red shirt as a symbol of open resistance and power, these units at times responded to reports of armed blacks. The Red Shirts, as they were called, intimidated and threatened black voters and disrupted Republican meetings. According to a Pennsylvania newspaper, the clear purpose of the rifle and saber clubs was to support the Democratic Party. A pro-Republican newspaper in New Orleans appreciated the extent of the growing threat. "Both Louisiana and South Carolina," the newspaper declared, "are afflicted with an epidemic among the Democrats which threatens to destroy the peace."[7]

There was a certain level of risk associated with the blatant use of bulldozer tactics. The Republicans could use accounts of paramilitary activity to stir anger among Northern voters and create a backlash against the Democrats. Indeed, the clash at Hamburg, South Carolina, was a case in point. A confrontation during a Fourth of July celebration between two white farmers driving a wagon through Hamburg and a parade of the town's all-black militia unit led to a legal complaint against the militia captain. Because the militia unit had been armed during the confrontation, it attracted the attention of white men in the surrounding area, many of whom belonged to local rifle clubs. Somewhere between 200 and 300 armed men swarmed the town on July 8, the day Matthew C. Butler, an attorney, recent candidate for lieutenant governor, and former Confederate major general, met with the trial justice hearing the case.

As the armed men assembled, Butler forgot his original purpose for being in Hamburg and quickly assumed the role of military commander. He ordered the militia company to turn in its arms. The militiamen occupied the building that served as their headquarters and refused to leave or give up its arms without orders from the proper authority. After negotiations to surrender the arms failed, shots were fired, and the militia returned fire, killing one man. This led to a full-scale battle, which included the employment of an artillery piece directed against the militia building. As the blacks fled the building, they were pursued, and a number were captured, including several who were not militiamen. Six men died, including the town marshal and five others taken as prisoners, who were executed shortly thereafter.[8]

The violence in Hamburg indicated the power of the rifle clubs to assemble rapidly and take violent action. Any threat of armed blacks, real or imagined, or any black state militia force displaying arms, would be met with a superior force willing to initiate battle and give no quarter. The Hamburg incident would have political repercussions but would only serve to highlight the failure of Northern will and the new political landscape that had emerged, both of which would shape the course of events during and after the election of 1876.

The new political landscape was characterized by the growing realization that "the constructive role of the Republican Party in the remaking of the South was nearly finished."[9] The Radicals, many of whom were either dead or no longer serving in Congress, were considerably weakened. The remaining Radical leaders in Congress were still as committed to preserve the strategic goals but had lost the ability to direct their strategy. Liberal Republicans, who had pulled away from supporting Radical efforts to pursue their Reconstruction strategy in 1872, were more concerned with tariff and civil service reform and good government, issues they believed were far more pressing matters that demanded the Republican leadership's attention. Carl Schurz, George Julian, and Charles Sumner were the most prominent of the former Radicals who embraced the liberal Republican program of turning to a new direction of compromise and conciliation. They realized that the goals of the Congressional Reconstruction strategy could not be accomplished without a significant increase in federal power and, concurrently, further government corruption. While still supporting the principles of the Fourteenth and Fifteenth Amendments that gave blacks civil and political rights, they advocated amnesty for all former Confederates and called for an end to federal forces propping up Republican state governments. Accepting white Southerner control of state governments in the South, they stressed, was the most rapid and surest means for a compromise to ensure honest government in the Southern states, while also avoiding future federal government involvement or interference in their affairs.[10]

Reconciliation and peace, along with prosperity for both races would be the happy result. The *New York Herald* reported as much in April of 1876, noting that under Democratic control, Georgia was enjoying prosperity, sectional animosities had declined, the laws were enforced impartially and effectively, differences in political opinions were tolerated, blacks were voting more independently, and a system of free education was in place. By 1876, the views of liberal Republicans had made deep inroads into the party and were to become the critical factor in shaping the approach to Reconstruction.[11]

A shift in perceptions and attitudes regarding blacks as political actors played an important role in swaying Republicans away from the original goals of Reconstruction. Prior to the war, slaves were viewed largely in the abstract—simply as human beings who were denied freedom. With victory, a new abstraction of the freedman emerged: citizens with civil and political rights, ready to become responsible independent workers and loyal Republican voters. But as blacks became more visible in American life, serving in Congress and in the Republican state governments, Northerners became less sanguine about the transition underway in the South. Freedmen were perceived as constituting a separate class, not altogether aligning with the Republican ideology.[12]

The inability of Republican state governments to establish true legitimacy or provide stability between 1871 and 1875 only heightened concerns in the North that blacks were not ready for the benefits of freedom. Blacks wanted the government to serve class interests to ensure social and economic advancement, rather than advancement through personal discipline, hard work, and high moral standards that the Republicans had envisioned would be the fruits of freedom and citizenship. The Southern employment of its asymmetric advantage on the information battlefield that portrayed the threat of the black militia to white Southerners fueled Northern fears of a black rural *lumpenproletariat* rising to destroy democracy—the counterpart to an increasingly restive class of Northern urban workers.

The image of an ignorant and depraved mass easily manipulated by corrupt and venal politicians was a powerful informational weapon that weakened Northern enthusiasm for any further efforts to uplift a people who seemed incapable of advancement, and increased sympathy for the argument that peace and stability in the South could only be guaranteed by white men. Northern newspapers were less willing to accept Republican claims that black voters were intimidated, pointing to information from Arkansas and Georgia that indicated blacks were turning away from "corrupt and rapacious carpet-baggers whose game is wholesale plunder and robbery of the negroes themselves" and voting for Democrats.[13]

This reformation of attitudes and perceptions in the North spelled doom for the Republican governments in Louisiana and South Carolina. Without the oversight and strong national arm of the Congress to support them, the state governments could only appeal to the president to employ military force to forestall their demise in the face of the revived Democrats, backed by the power of the bulldozers, who intended to control the source of the Republican state's power in the upcoming election. "The radical leaders know that the day they lose control of the Southern negro vote, seals their doom forever in the South," a Louisiana newspaper observed. "The republican party is responsible—alone responsible for the impoverished conditions in the South—for the massacres, riots, robberies and lawlessness rife in these States for the last eleven years, and it cannot escape the responsibility."[14]

Sen. George S. Boutwell of Massachusetts, a prominent abolitionist, who headed the Senate committee investigating the Mississippi elections of 1875, maintained that the Congressional strategy was still justified. The Congressional Republican Reconstruction strategic goal was to elevate slaves to the status of citizens and ensure their constitutional rights as citizens. Boutwell asserted that the only mistake the Republicans made in their strategy was displaying any magnanimity to the defeated Confederates. "We vainly imagined that gratitude to a common benefactor for a common benefit would lead them to guard the rights of the negro as they would desire and expect the country to guard theirs." As Republican Reconstruction state governments fell to Democratic control, "and held by fraud and force," Boutwell called attention to a new threat. "A large part of the able-bodied white men of the Gulf States," he reported, "are completely armed and enrolled in military organizations, whose designs are aggressive and purely political."[15]

During the summer of 1876, President Grant received a significant amount of correspondence from citizens who portrayed the South as a region in crisis. One traveler, who had visited Georgia, Mississippi, Alabama, and Louisiana, among other Southern states, reported that "seeing and hearing the people talk," it was his belief that there was an immediate need for troops to be deployed before the upcoming election, "to protect the large body of voters in their rights." A woman wrote from Texas describing the murder of her husband and told the president bluntly that "Union Men and Woman will be either killed or drove [sic] out of the State."[16] A writer from Natchez, Mississippi, stated that "the Republicans are not permitted to assemble for fear of violence," and appealed to the president to send federal troops, "that we may be protected and saved from personal destruction." Another letter from Louisiana reported that the Democrats had a universal hatred of Republicans, and they were organized in such a way that they resembled the Confederate army. With an astute appreciation for the president's priorities, the writer claimed that every black man in the state would vote Republican if allowed and that federal troops were essential for the Republican vote to come out.[17]

In August of 1876, bulldozer tactics were directed against South Carolina Republican governor Daniel H. Chamberlain, who was speaking at Edgefield Court House—one of the strongholds of white Southern resistance. A group of 600 armed men arrived and demanded that Democratic Party speakers be given half of the allocated speaking time. Governor Chamberlain spoke first, but, it was reported, "under great constraint." He was followed by Gen. Matthew C. Butler "of Hamburg fame" and several others who denounced Gov. Chamberlain and his Republican supporters, described as "chiefly negroes." Shortly afterward, they were followed out of town by a jeering mob.[18]

President Grant was stirred to action, decrying the Hamburg riot and other instances of violence, and issued a proclamation commanding the rifle clubs and other armed organizations in South Carolina to disperse within three days. In addition, he instructed the secretary of war to order general-in-chief of the army William T. Sherman to hold in readiness all the available forces not engaged in Indian warfare to be used to protect the rights of all citizens to vote and to prevent threats to voters in the form of violence or intimidation.[19] The proclamation, however, had the opposite effect of its intent, indicating the weakening will of the North and faith in the motives of the administration. To many observers, President Grant was using the military as a screen for ensuring a Republican electoral victory. "The election in the Southern States are to be carried by the bayonet," claimed a Pennsylvania newspaper. "For every vote in the South restrained by the presence of Federal bayonets at the polls," thundered a Delaware newspaper editorial, "a thousand votes in the North [will] be cast on the side of justice and liberty, and for the rights of the people. Let the Administration beware, for the day of reckoning is at hand."[20]

The day of reckoning appeared in South Carolina in the form of Wade Hampton, who had been nominated in September by the Democrats as the candidate for governor. Hampton was the state's greatest hero of the war, a dashing cavalryman, the commander of Hampton's Legion, who had fought in every major battle of the Army of Northern Virginia, eventually serving as Lee's cavalry corps commander. During an early campaign swing through the up-country, Hampton stressed reconciliation, home rule, and good government, but everywhere he went he was accompanied by Red Shirts, armed horsemen whose numbers at times reached as large as 1,500 men. The bulldozer activity continued to target Republicans meetings and intimidate voters; a number of bloody riots occurred, reflecting the tensions and fears within the state. Hampton discouraged any violence but endorsed the demonstration of force of the mobilized Southern identity, now fully and openly on display during the campaign.[21] This was an important aspect of bulldozer tactics employed as asymmetric warfare. As an indirect approach, it was a demonstration of overwhelming power that was intended to sway blacks to vote for Democrats or at least dissuade them from risking a vote for Republicans. Behind this display of military organization and firepower lay the implied threat that in the face of any provocation, these units were capable of exploding into extreme violence that could escalate into annihilation.

In early October, Governor Chamberlain decided to forestall the growing momentum and counter the growing effectiveness of the new asymmetric tactics by ordering "unlawful combinations and assemblies of rifle clubs and all other organizations or combinations of men or formations to disband and cease to exist."[22] President Grant provided the threat of federal intervention with a proclamation that associated the rifle clubs with insurrection, declaring them "combinations of men against law," and ordered those involved to return to their homes.[23] The president then issued the action order for his alert order that had been sent to the War Department the previous month. Grant passed instructions to the secretary of war to order General Sherman to dispatch all the available units in the Division of the Atlantic to reinforce Gen. Thomas H. Ruger, commander of the Department of the South at Columbia, South Carolina. Ruger was ordered to station troops in localities where resistance to federal authority was most anticipated. With only about 2,800 soldiers available, Ruger was limited in what he could actually accomplish. One infantry company and four artillery batteries in Virginia were alerted for movement to reinforce Ruger. In turn, Ruger issued orders that the 19 companies in South Carolina and the scattered detachments in Florida would be placed close to polling places, but they were directed that unless there was a request for support from a federal officer, they could not attempt to interfere. General Augur, commander of the Department of the Gulf (which included Louisiana), would follow suit, eventually deployed his detachments in 62 locations.[24]

The Ohio *Gallipolis Journal* took an unexpectedly hard line in favor of the proclamation declaring, "General Grant's orders will be obeyed, and it may be necessary, as it was in 1861, to kill a few more Democrats that a free ballot, and a free government, may exist." A few days earlier, another Ohio newspaper had the exact opposite reaction, announcing "Bayonet Rule," and displaying the headline: "South Carolina to be Overrun with Federal Soldiers," followed by two even more provocative subheadlines: "The Napoleonic Plan of Bayonet Voting" and "Once Established in the South it Will be Tried in the North."[25]

After Grant's proclamation and the announcement of federal troops moving to South Carolina, the State Democratic Executive Committee quickly published what it called an address to the people of South Carolina. It was obviously crafted for a far wider audience as it declared the essential arguments of the mobilized collective identity. According to the committee, Chamberlain's dispersal order and the federal response represented nothing more than the arbitrary use of force against a people who had been suffering under a dishonest and dishonorable Republican administration. The committee claimed that the rifle clubs and associations had been formed for home protection against state politicians who were agitating and encouraging blacks to employ violence and arson against whites. The committee asserted that army units arriving in the state will find "no insurrection against the government, no resistance to civil process, and the whole country at peace except where the colored militia, armed with State guns, are assaulting and killing the whites."[26]

The *New York Tribune* appeared to agree with the Executive Committee's address, noting in an editorial that "General Grant's administration has so often interfered in Southern affairs pending elections, and almost always on behalf of the plunderers who have disgraced the name of Republicanism." It concluded that "people have learned to look with suspicion on the use of United States troops in any southern state during a political contest," and noted that earlier deployments of federal troops had led to political backlash in the congressional elections two years earlier.[27]

The effectiveness of the asymmetric strategy to delegitimize the Republican Reconstruction state governments was on full display. The president's final attempt to bolster the Congressional Republican strategy had little effect. Grant's actions were widely derided in the Northern press as a feckless political display, while within South Carolina, the arriving troops were met with cheers and applause from the people. These demonstrations of goodwill were intended to convey a message that no threat existed and that Chamberlain was creating a false crisis, both to stir Northern sympathy and to establish the plausible pretext for the president to come to his aid.[28] A North Carolina newspaper reflected the now predominant view that the Republicans controlling the state governments in the South owed their strength solely "to the countenance received from the national government." Whenever the Republican governors were in trouble, "the bayonets of the Federal army have been summoned . . . and they have come as promptly as they were required."[29]

The dominant actor attempted once more to strike a blow to defeat the weaker actor's efforts, but even as action was taken, the weaker actor's dominance of the information battlefield demonstrated that Northern will to support further military action had faded, and the weaker actor's sophisticated demonstration of political-military effectiveness would largely negate the federal forces deployed to protect the black voter. In fact, in all the Gulf States, but predominantly in Louisiana and South Carolina, black voters were assured if they supported the Democratic party, not only would they have physical protection from intimidation, but their full civil rights as citizens and voters would be recognized and respected. These inducements undoubtedly had some effect and supported the claim of white Southerners that blacks were independent voters who were less than satisfied with the Republican state governments and willing to cast a vote for Democrats.

Rutherford B. Hayes, the Republican nominee for president, was a former congressman who had voted for the Civil Rights Act of 1866 and was known as a supporter of the rights of freedmen. In his July 8 letter accepting the nomination, Hayes recognized that establishing peace in the South based on the rule of law was paramount. He made it clear that he sought reconciliation through the promotion of "political and private rights" for all citizens through "honest and capable" state governments. As president, he would seek to pursue a civil policy that would forever eliminate any distinction between North and South. These conciliatory words were intended less for Southerners and more for the Liberal Republicans, and they masked a real fear that losing the election would turn the national government over to former Confederates, which meant that the Union's victory in a war for what Republican James A. Garfield called "equal and universal freedom" would be rendered meaningless.[30] The *New York Times* put the issue quite plainly:

> The choice is not simply between one party and another, it is between the principles which assailed the Union and the principles which saved it. Between men who were rebels and who were loyal and true; between the policy that regards reconstruction as an accomplished fact and the policy that would undo reconstruction and invest its enemies with power.[31]

Samuel J. Tilden was a lifelong Democrat from New York who worked tirelessly to support the party. A wealthy lawyer, he spent the war managing the senatorial and presidential campaigns of Horatio Seymour. By 1874, he had played a pivotal role in the breakup of the Tweed Ring in New York City and was elected governor. As the Democratic nominee for the presidency, Tilden could likely deliver his state's electoral votes, which were essential to breaking the Republican stranglehold on the Electoral College. Equally important, with his reputation as a reformer, he was attractive to Liberal Republicans, who might be induced to cross party lines. In his letter accepting the nomination, Tilden focused on economic issues, relating the troubles in the South to Republican economic mismanagement as well as misgovernment, while supporting the civil rights of all citizens.[32] Although actively courting Southerners, Tilden distanced himself from the bulldozer tactics and the episodic violence in South Carolina and Louisiana that he feared only highlighted the Republican national chairman Zachariah Chandler's accusations of "Southern barbarism, ostracism, and persecution," and the widespread belief within the party that a Democratic victory would turn the nation over to the authors of rebellion.[33] Although white Southerners were resentful of Tilden's equivocation, there was no doubt that their votes belonged to the New Yorker. From the perspective of the asymmetric actor, national political support meant less than tactical success in assuring the fall of the Republican state governments.

The Grant administration was confronted with a significant barrier to action in the South as public sentiment against the Republican Reconstruction goals became more strident during the election season. A West Virginia newspaper went as far as declaring the Republican Party dead because it was still bound to "issues that gave it being and which have long since passed into history." At the Union Soldiers' and Sailors' Reform Association, meeting in Washington D.C., members denounced the "dictatorial spirit of Radical leaders," whose only purpose was "to rekindle the old war feeling of animosity." The veterans made it clear that the Republicans could no longer consider their votes guaranteed as long as the party continued "to denounce the people they conquered" while at the same time "perpetuating Radical Republican misrule."[34]

A Kentucky newspaper made the argument that the Republican Party's pursuit of its goals in the South had led to a loss of 10 years of economic prosperity for the states of Indiana, Ohio, and Kentucky due to the corruption of the "radical-ridden" Republican Reconstruction state governments and the high taxes imposed on the Southern people "by carpet-bag thieves." The violence that had engulfed the South over the decade was nothing more than clear "evidence of incapacity or oppression of those in power." Peace, quiet, and prosperity could only return to the South with the elimination of Republican control.[35]

For many observers, the election did not offer significant contrasts between the Republicans and Democrats. An Ohio newspaper noted that both parties "declare in the same terms for honesty and economy of administration and holding officers to rigid responsibility." On the Southern question, both parties proposed "a just constitutional policy." The difference was that the Democrats accepted the Fourteenth and Fifteenth Amendments as the resolution of the outcome of the war, while the Republicans emphasized proper enforcement of the amendments. The paper concluded that with no significant differences between the two parties, "the election will be decided to a great degree by personal considerations." The *New Orleans Republican* had a far more serious prediction, based on the growing armed presence in Louisiana. If Tilden did not win the election, "the land would be deluged in blood."[36]

In Louisiana, the Democratic nominee for governor was Francis R. T. Nicholls, who, even though a former Confederate general, had gained a considerable following among whites as well as blacks. Nicholls astutely appealed to the mutual interests of black and white voters and campaigned actively for black votes. His opponent was Stephen B. Packard, one of the most notorious members of a corruption ring exposed for siphoning off millions of dollars in tax revenue, antagonizing voters of both races. Packard took the black vote for granted and expected victory. The main threat to Packard was the looming presence of the White League, fully armed and prepared for action in short notice. Its goal was to "impresses the blacks with united strength," and "to wrest the government of the state of Louisiana from an alien band of robbers and restore it once more to the hands of her own people," by eliminating "the effects produced by ignorance and vice."[37]

The election proceeded with little overt violence, but it can hardly be described as fair and free. Because of the high political stakes, the bulldozers in South Carolina and Louisiana were active and, to a certain extent, successful. In Edgefield, South Carolina, the ballot boxes designated for black voters were surrounded by heavily armed Red Shirts mounted on horses or mules wearing some modification of a Confederate uniform. A company of U.S. infantry was needed to open a corridor to allow the few brave individuals to walk the gauntlet of grim-faced men pointing weapons at them.[38] In Ouachita Parish, Louisiana, about 200 bulldozers armed with Winchester rifles, shotguns, and Colt six-shooters occupied Monroe. Although a detachment of federal troops was present, the bulldozers controlled access to the ballot boxes. A black deputy U.S. marshal was attacked and seriously wounded as he transported a ballot box in the parish.[39]

An alternative and very influential voice came from W. G. Summer, a professor of political and social science at Yale, who was one of the founders of the science of sociology in the United States. Shortly after the election, he wrote a widely published and influential article on political and social conditions in Louisiana. Summer found that whites were not fomenting disunion or were necessarily hostile to black suffrage. What whites desired was what they called redemption, which would serve as the basis for racial peace: "They thought that if they could once get self-government for the people of the State, they could manage to live together without trouble." Summer believed the goal of self-rule, or home rule as it was also known, was justified. He characterized the Republican government in Louisiana as a "prodigious tyranny," marked by "maladministrations of the civil offices, fraud, chicanery and abuse." His conclusion was nothing less than a realization of the dominant actor's fatal assumption that it could impose its will and achieve the pacification of the Southern states. For Summer, conditions in Louisiana only proved "how wrong it is for us to be governing a state sixteen hundred miles away."[40]

Shortly after the election, General Sherman received instructions from President Grant to inform General Ruger (commander of the Department of the South with responsibility for Florida and South Carolina) and General Augur (commander of the Department of the Gulf with responsibility for Louisiana) to ensure that the forces assigned to their respective commands preserved order and to report any instances of fraud in counting the returns. Grant's concerns were justified. The potential for extreme reactions to a disputed election were real. On November 17, W. Jasper Blackburn, a former Republican congressman who was serving as a Louisiana state senator at the time, reflected the attitude of many extremists when he declared that if the results of the election were contested, "then let us have another war and an electoral separation, or another policy of reconstruction." A Texas newspaper expressed a surprisingly different point of view. "If a conflict of arms must ensue," it laconically observed, "there is no need for our [the state of Texas] intervention." Northerners will rise up on their own, and then, "the whole people will array themselves on the side of honesty and right and the army and police can hardly resist."[41]

In the aftermath of an election that had the largest voter turnout in U.S. history and saw white voters in the South participating in unprecedented numbers, Florida, Louisiana, and South Carolina declared their electoral votes for Hayes; the Democrats protested that Republican-dominated return boards had skewed the count in favor of Hayes by invalidating Democratic votes. It was clear that the Republican intent in these three states was to certify Hayes as the winner of the election and deny Tilden their electoral votes. In addition to the disputed electoral votes, the results of the gubernatorial contests were also in doubt, leading to dual legislatures and dual governors in the capitals of South Carolina and Louisiana, each claiming authority to govern.[42]

Establishing a parallel government served as an especially effective tactic of asymmetric warfare. Because this election was so crucial to the survival of the Congressional Reconstruction strategy and at the same time represented the final assault by the weaker actor to force the dominant actor to accept a new political status quo, any contested election result that went against the Democrats in either South Carolina or Louisiana would result in the losing candidate refusing to accept the legitimacy of the election while claiming the authority to establish a legal government. This shadow government, implicitly backed by armed force, could remain in place while the results were adjudicated. This process took so long that the time advantage went to the government that could maintain its power, while uniting popular opposition against the rival government in order to weaken its claim to legitimacy.

In South Carolina, there were 300 U.S. troops in Columbia ostensibly keeping order, but they were surrounded by 5,000 Red Shirts with repeating rifles, who were described by a New York newspaper as "regularly trained military organizations, made up in great part of the veteran soldiers of the Confederate army." The correspondent described a condition of "great strain of anxiety and apprehension lest scenes of violence and bloodshed should set the whole country on fire and inaugurate a new civil war." Hampton pledged that there would be no resort to violence, and he promised to repress any use of exhibition of force. Command of the city had defaulted to Hampton and his Red Shirts.[43]

In Florida, federal troops were deployed to predicted danger areas such as Marianna in Jackson County as well as the cities of Tallahassee, St. Augustine, Pensacola, Gainesville, and Tampa. Voting irregularities were committed by both sides, but there was no direct violence. After the election, 12 companies of federal troops arrived in the state capital at Tallahassee, ostensibly to maintain order and ward off any potential armed incursion in the aftermath of the disputed election. They remained mostly bystanders, although several small detachments were sent to Jackson County to review election procedures. Gov. Marcellus Stearns, who ran for reelection, had initially been declared the winner over Democrat George F. Drew, but thanks to the intervention of the state supreme court, Drew was instead declared the winner of the election and inaugurated.[44]

In the midst of the turmoil and danger in two critical states, Congress took on the challenge of determining the next president. Congress formed an electoral commission representing members of the House, Senate, and Supreme Court to validate the electoral count from the certificates from each state that had been sent to the president of the Senate. The commission had a little over a month to complete its work. A decision had to be made by March 4, the day the new president would be inaugurated. After much debate, and by a partisan vote of 8–7, the commission accepted the Florida, Louisiana, and South Carolina electoral votes for Hayes. As it became clearer that Hayes would be certified, a group of prominent Republican congressmen met with important Southern Democratic congressmen in late February to initiate what would become the process of establishing a political equilibrium and an acceptance of the new status quo brought about by the weaker actor's employment of asymmetric warfare. The Republicans agreed to shape Hayes's administration policy regarding Reconstruction in the South, accepting Democratic control of Louisiana and South Carolina, in exchange for an acceptance of the commission's electoral count and a pledge that the prospective Democratic governors would respect the constitutional rights of blacks. After a day and night of turmoil in the House of Representatives, which included Southern and Northern members squaring off for fisticuffs or drawing weapons on one another, the president of the Senate declared Rutherford B. Hayes the president-elect just two days before the March 4 deadline. Hayes, who had arrived in Washington that day, was secretly sworn in on March 3 to ensure a peaceful transition to power.[45]

Hayes took his time in finalizing the withdrawal of federal troops from Louisiana and South Carolina, which marked the establishment of political equilibrium. In his augural address, Hayes made it clear that he expected the Sothern state governments to acknowledge their obligation to accept and support the maintenance of constitutional rights for all citizens. But he also understood that the consensus that had emerged to create this movement to equilibrium had helped him become president. The House of Representatives signaled its intent to force the president's hand when it blocked the annual military appropriations bill that would continue to fund the army, including those units deployed in the South. This action, in effect, guaranteed that the president would have no immediate capability to deploy U.S. troops to the South again even if he chose to do so.

In late March, Hayes took on the least difficult challenge, summoning Daniel Chamberlain and Wade Hampton to the White House to hear out each claimant to the governor's seat. Hampton assured Hayes that he would ensure the equal protection of both races. In a letter to Hayes, sent on March 29, Hampton stated that, "the rights of all shall be safe and the interests of all shall be protected." On April 3, the president issued the requisite order to withdraw the federal forces that were in Columbia keeping the peace between the rival factions by April 10. On that appointed day, 20 U.S. soldiers under command of a lieutenant departed the statehouse precisely at noon. The following day, Hampton became the governor of South Carolina.[46] The withdrawal of federal troops was described by a South Carolina newspaper as "an auspicious event for which the whole country should thank President Hayes and Governor Hampton and his fellow citizens." The *New York Herald* acidly commented that Governor Chamberlain, even with the backing of federal troops, "was powerless to preserve order, powerless to repress violence, protect life or punish crime."[47]

In Louisiana, the situation was more volatile. Anticipating trouble in New Orleans over the impending vote count, General Augur assembled almost three infantry regiments—nearly 1,200 men—in the city and stationed two gunboats at the docks. Governor Kellogg sent a request to the secretary of war asking for General Sheridan to return to New Orleans to demonstrate the government's support for the Republicans and deter any potential violence during the vote count. Sheridan did arrive, but reluctantly, and unlike his previous visit, marked by bluster and bravado, he kept a very low profile. Sheridan had no intent of taking command from Augur, nor was he modifying the orders of General Sherman, who had restricted Augur to employ the forces under his command only to keep the peace in case of violence. In fact, Sheridan was quite satisfied with everything Augur had done and departed the city soon afterward, confident in the widely held foregone conclusion that the all-Republican returning board would certify Rutherford Hayes's election as president and Stephen B. Packard's election as governor.[48]

With the announcement of the returns, John McEnery, who still maintained that he was the legitimate Democratic governor from the 1872 election, joined with gubernatorial candidate Francis R. T. Nicholls in refusing to accept the decision of the returning board. Nicholls and the Democrats would establish a parallel government but with a show of force that would leave Packard with no option but to surrender his claim. On January 8, each governor took the oath of office. On the following day, when the Republican state legislature was scheduled to convene, 3,000 armed men, now designated as the state's legal militia, under command of Fred Ogden, who had led the White League in the street battles of 1873 and 1874, took control of government buildings, the headquarters of the Metropolitan Police, and the state arsenal. Unlike 1874, neither federal troops nor the Metropolitans displayed any indication to contest this show of force. Because there was no violence, Augur kept strictly to his orders and took no action. The militia surrounded the statehouse, isolating Packard and the Republican legislature, and the members of the state supreme court were ejected and replaced with pro-Nicholls justices.[49]

Military control of the city allowed Nicholls to establish a functioning shadow government, with a structure of rotating units to provide security and financial support sustained through voluntary donations and other resources, such as the Louisiana Lottery Company. Packard appealed to Augur for assistance, but he refused after being notified that President Grant needed time to review the situation and determine which government would be recognized as legitimate. General Sherman fully supported Augur's execution of his orders by avoiding any involvement except to suppress violence. Although Grant indicated that Packard was the legitimate governor by the returning board's decision, he also was clear in signaling that he had no intention of influencing events with military force. Grant ultimately had indicated that the federal government no longer intended to pursue its Reconstruction strategy. In the face of a determined political movement backed by a powerful armed resistance, the cost of sustaining the Republican Reconstruction government was too high—the government would have little or no role in resolving the Southern problem. The *New Orleans Republican* admitted defeat, recognizing that U.S. troops over the years had been unable to protect black voters from violence or intimidation either before or after an election. The black Republican could vote only through "the toleration of the people."[50]

As the drama of the electoral vote count continued over the next few months, it became clear that Grant intended to leave the situation frozen until the inauguration of President-Elect Hayes. Quite simply, the dominant actor was signaling an acceptance of an outcome far short of the desired strategic goals of Congress. Hayes, in turn, also waited, taking deliberate steps to employ the thin screen of a presumed democratic process by appointing a presidential commission to investigate the situation in Louisiana. The commission had no other agenda except to find an expedient solution by bargaining to combine the two legislatures, which would give Nicholls a majority and make him the governor by default. But the time for the Packard faction in the legislature to accept the offer to switch sides was slipping away quickly. The president was already taking action to withdraw troops from the statehouse, intending to leave the Packard government to its fate. After gaining further assurances of protection of the rights of blacks, Hayes passed the order on April 20 to remove the five infantry companies guarding the statehouse, where they had maintained a symbolic presence between the contesting legislatures. On April 24, the soldiers in full dress paraded from the Orleans Hotel where they were stationed and loaded transports as a vast crowd watched and cheered the victory. The *Chicago Times* correspondent who witnessed the event proclaimed that the departing soldiers were "once more the chosen defenders of the nation, and not the mercenary jailers of a section of the great republic."[51] Packard, recognizing the inevitable, simply relinquished the governorship to Nicholls, who took office officially on April 26.[52]

Yale professor Leonard Bacon, who prior to the war had been active in both the temperance and antislavery movements and was widely considered the most prominent Congregationalist of his time, published a letter in the *New York Times* that seemed to represent the sentiment of the majority of Northerners as the drama of the electoral count continued. The Congressional Reconstruction goals must become subordinate to larger Republican Party interests. It was time to "consult and cooperate with all those who will unite to give the recovered States the best practicable government in the earliest practicable way." Bacon accepted the most basic requirement for the establishment of equilibrium, which he described as "all men have civil and political rights but a limited franchise." He believed that with the ballot alone, "freedmen would be able to defend themselves and permanently strengthen the Republican party."[53]

The redoubtable Indiana senator Oliver P. Morton gave a bitter assessment of the effectiveness of asymmetric warfare, the significance of the information battlefield, and the influence of a mobilized collective identity at the symbolic end of Reconstruction in May of 1877. The Republican governments of Louisiana and South Carolina had yielded to "an armed minority whose threats of future violence were guaranteed by a long train of bloody deeds in the past." The Republican Party lost Mississippi and Alabama through a combination of violence and criminal activity. "We have had so much talk about conciliation and fraternity," Morton observed, "that many well-meaning people in the North have come to believe all that was necessary to secure tranquility and equal rights in the South was to withdraw the army." The political violence in the South had often been obscured or rendered insignificant by the Southern claim that the targets were "carpet-baggers, scalawags and radical thieves." In the end, it was the unified action of white Southerners that frustrated the efforts of the dominant actor. "The large body of the white people who engaged in the rebellion," Morton concluded, "are firmly united in favor of several things, and they will stand by the party that favors them and oppose to the bitter end the party that opposes them."[54]

With the news of the withdrawal from New Orleans, 100 guns were fired in Shreveport in salute; in Canton, Mississippi, 38 guns were fired in salute to Louisiana; Meridian, Mississippi, fired 138 guns; and West Point, Mississippi, fired 25 guns. Charleston, South Carolina, fired 50 guns; Columbia, South Carolina, fired 17 guns; and Mobile, Alabama, fired 100 guns. From Texas, Gov. R. B. Hubbard sent his personal congratulations to the people of Louisiana celebrating "the removal of Federal bayonets."[55] It was a triumph of the mobilized collective identity of the white South. As one newspaper reported:

> Men and women and Christian clergymen are giving thanks to the Supreme being for having overthrown what they are pleased to call the thieving Radicals, the cowardly carpet-baggers, the corrupt negroes, the low politicians, who have robbed them and disgraced their country.[56]

Southern virtue and honor had been restored; Confederate unity and courage in the face of overwhelming odds again had been validated, and the methods of warfare they had initiated to bring about this victory had been justified. A letter from an "old farmer" to a South Carolina newspaper captured the spirit of triumph. "By a desperate struggle," the Red Shirts had driven out "the horde of robbers that have for years plundered our people." It was now the time for "vigorous, active, ambitious" young men "to assist us in restoring our State to her ante bellum prestige." In the words of a North Carolina newspaper, the "self-governing, self-redeeming" Southern states had finally become quiet and orderly, "ruled with honesty, justice, and fidelity to the rights of the people and the interests of the nation." A Cairo, Illinois, newspaper declared "carpetbaggery is at an end," and with its demise, "all race troubles between whites and blacks in the south."[57]

These sentiments were contrasted by a Mississippi black leader in an October 1875 address that recognized the future of blacks as political actors that would come about as the result of the new political equilibrium in the South. The black voter was "free in name but not in fact—poor, ignorant, helpless," he declared, "edged in by unfriendly laws which he will have no power to circumvent."[58]

Because the Democratic governments inherited the Republican power structure that still would allow the Republicans (primarily black voters) the opportunity to challenge the status quo, new constraints had to be put in place. Although blacks would have the same rights as in the North, such as the right to work for themselves, to make their way forward the best they could, and vote if they chose to do so, they would never again, in the words of one Mississippi Republican, "be permitted to ride rough shod over the educated classes and property interests." Rejecting the Reconstruction electoral system as illegitimate, white Southerners took action to make the system ineffectual and keep the Republican voters from having any immediate or future influence in state politics. While control of the individual black voter was the focus of the effort, violence was no longer needed;

instead, a combination of legal and extralegal activities accomplished this intent.[59]

It is commonly accepted that the compromise of 1877 marked a major milestone in U.S. history. Simply put, Republicans retained the presidency in return for withdrawing federal troops, thus ending Reconstruction. In actuality, what emerged in the aftermath of the extraordinary election of 1876 was not the end of Reconstruction but a political equilibrium that ended the South's employment of asymmetric warfare against the federal government. This equilibrium was represented by political compromise and a mutually preferred settlement involving "signals of conciliatory intent." Equilibrium resulted from a process that involved arranging credible guarantees through concessions obvious to both parties that represented the necessary costs that each side was willing to bear in order to establish the new norms of peaceful cooperation within a mutually acceptable domestic order.[60]

Equilibrium had been achieved because President Hayes and the Republican administration that he intended to form was a necessary cost acceptable to Southerners with his signaled intent to relinquish the pursuit of the goals of the Congressional Reconstruction strategy by withdrawing the forces that sustained the last vestiges of Republican control in the South. Moreover, Hayes represented a counterpoise to the Radical element in the Republican Party and a welcome proponent of good government in the aftermath of the roiling controversies of corruption that characterized the Grant administration. In return, however, Southerners conceded to the necessary costs of Republican national dominance and an acceptance of the Republican ideology directing the future of the nation, based on what the *New York Herald* described as "the political, personal, social and commercial restoration of the Union."[61]

For the Republicans in the election of 1876, one immediate political objective became predominant: in order to secure the presidency and protect it from a Southern influence still believed to be tinged with treason, the dominant actor accepted the necessary costs of the demise of the Republican Reconstruction governments in the South. The Grant administration had already signaled that the federal government was no longer willing to bear the costs of defeating an asymmetric opponent to achieve the goals of the Congressional Reconstruction strategy. By taking the symbolic act of withdrawing U.S. troops from their role as political peace enforcers, Hayes was more or less conceding the existing conditions of a political realignment in the South. In return, he was able to secure the minimally acceptable strategic goal of the Republican Reconstruction strategy: a nominal guarantee of the protection of constitutional rights for all citizens. This assurance came from the new governors of South Carolina and Louisiana, both of whom had been prominent Confederate generals. Hampton had bolstered his election prospects in South Carolina through the overt demonstration of armed force; Nicholls had achieved political control in Louisiana by the employment of armed force that, for all intents and purposes, amounted to a military coup. The asymmetric power the weaker actor had brought to bear over 10 years against the dominant actor led to the mutually acceptable domestic order that both the North and South had sought since 1867. This equilibrium marked, in the words of one historian, "the Republican administration's concession to the privilege of peacefully retaining power."[62]

The U.S. attorney for West Texas, in a March 1876 dispatch to Atty. Gen. Edwards Pierrepont in Washington, perceptively described the purveyors of asymmetric warfare in his state as "educated law-breakers, educated disorderists, educated resisters of law and order."[63] One of the most critical aspects of asymmetric warfare is the weaker actor's knowledge of the dominant actor's means and methods. One reason a weaker actor resorts to asymmetric warfare methods is to create a series of dynamic conditions that influence the dominant actor's desire to attain its goals. By maintaining a level of disorder and operating within certain limits, the asymmetric actor can act with impunity and frustrate every effort of the dominant actor to suppress the resistance and wear down the opponent's will. If the weaker actor is successful in its pursuit, the dominant actor will be forced to recognize the sources of the resistance as beyond its power to affect and work toward a negotiated settlement.[64] In Phase III, the Southern "educated law breakers, resisters, and disorderists" had succeeded in gaining every advantage over federal and state power. From 1871 to 1875, violence served to displace freedmen and loyal whites and changed the political survivability the Republican state governments. The battles in Phase III increasingly moved into the political arena, with the intent of developing political momentum characterized by open defiance of the Republican government through an often openly armed popular movement. Whenever federal or state power was used, the resisters and disorderists disappeared, only to reemerge in a different form, but stronger and more organized. From the success of this approach, the bulldozing tactics and overt military operations in Louisiana and South Carolina emerged in the last Republican strongholds in the South.

In Phase III, the state governments fell one by one to the Democrats, who promised the peace, stability, and progress the Republican governments appeared unable to deliver. In two significant aspects of asymmetric warfare, the weaker actor demonstrated the asymmetric advantage of will and dominated the information battlefield in such a way that the dominant actor was never able to match. The Northern will to bear the costs of sustaining the Congressional Reconstruction strategy continually faded as public opinion increasingly supported the Southern depiction of the Republican Reconstruction governments. White Southerners demonstrated a willingness to escalate armed confrontations to a level that neither the state nor the federal authorities were willing to challenge without accepting the prospect of a renewal of open hostilities and a race war.

In the new political environment that arose in the aftermath of the state and national elections of 1876, it became clear that some form of accommodation was necessary. The North would have to accept the current conditions in the South or commit itself to a new Reconstruction strategy that it had neither the will nor the capability to pursue.

The course of asymmetric warfare in 1876 served its purpose for Southerners. It revealed to the North the relative strength and determination of the resistance, both violent and nonviolent. It became clear to the resistance that the will of the North to pursue its preferred outcome in the South would limit the options and actions of the Grant administration. The series of events in South Carolina and Louisiana presented the North with the choice of continuing to pursue its strategic goals or consider some sort of bargained settlement. The election of 1876 and its disputed outcome revealed in words and actions this understanding and offered the two sides the opportunity to bring about some type of political equilibrium.[65]

As the effectiveness of federal military or legal action faded, along with the will of the population of the North to sustain the effort necessary to achieve the goals, the only option left to the dominant actor was a political solution that would achieve some measure of strategic success without further cost. It was a tragic irony—after years of strife, enmity, and violence, the political equilibrium that emerged in 1877 very closely resembled the original, but short-lived equilibrium of 1866.

9

Asymmetric Warfare, Reconciliation, and the Road to the New South

The political equilibrium that manifested itself during the first few months of President Hayes's administration illustrated how the weak actor's asymmetric advantage in demonstrating a far greater level of resolve and will made the dominant actor's power irrelevant and led to a political accommodation that no longer required military force to sustain. By returning South Carolina and Louisiana to Democratic state control, a political equilibrium emerged that was acceptable to both actors. Both actors had achieved their essential and most important requirements to satisfy their goals. The dominant actor gained civil peace and a guarantee of respect for the essential constitutional rights for black Southerners; the weaker actor gained regional political control and autonomy along with access to federal aid.

Yet this equilibrium came with no real guarantee of permanence. The protracted period of asymmetric warfare had been characterized by a steady state of violence, unnumbered thousands of deaths and injuries, and great anger, bitterness, and hostility. The Southern-Confederate collective identity had been mobilized to deal with the threat of the freedmen as a political actor and to delegitimize the Republican Reconstruction state governments established by the orders of Congress. The North, as a dominant actor, was still perceived as a menacing enemy, ready to reimpose its power over the South. For the North, the image of the South as a barbarous, violent, and malevolent society far removed from the American mainstream brought to question the sincerity of Southern guarantees of black civil rights.[1]

Although asymmetric warfare had driven the dominant actor to accept a political equilibrium, there were a number of barriers to long-term peace in the aftermath of a long and arduous social, political, and psychological struggle. Southerners and Northerners would have to overcome these barriers and establish a long-term peaceful relationship marked by reconciliation, which, in essence, is characterized by the restoration of "friendship and harmony between rival sides after conflict resolution." It is a mutual and consensual process of recognition and acceptance, involving all domains and all sectors of society changing attitudes from supporting conflict to favoring peaceful relations.[2]

This process of reconciliation requires patterns of cooperative interaction and interdependence in political, economic, and cultural relations. It also requires a psychological element—a willingness to redefine the relationship—involving changes in motivations, actions, perspectives, attitudes, and emotions. New goals for the society must be defined, and new frames of identity must emerge, supported by conciliatory acts and a common outlook on the past. These goals and new identity frames must be used to communicate the change in the relationship to replace mistrust and hatred with mutual trust and respect. In this crucial process of reconciliation, public ceremonies involving rituals and symbols mark the transition of former enemies to brothers who have resolved their former differences.[3]

Most importantly, reconciliation involves establishing a mutual respect for each other's collective identity. Changing collective identity requires "a degree of acceptance of the other's identity—at least in the sense of acknowledging the legitimacy of the other's narrative without necessarily agreeing with that narrative." This revision of the collective identity is possible only if the core of the identity remains intact. Identity is often closely linked with dignity and self-esteem. Northern recognition of the Southern-Confederate collective identity would restore a sense of dignity and self-esteem to white Southerners that had been lacking since Appomattox. In outlining these required changes in society, it is not surprising that the process of reconciliation lasts many years, decades, or even centuries.[4]

In the fall of 1877, President Rutherford B. Hayes sought to bring Southerners into the national fold by emphasizing nonsectional and nonracial issues related to reinforcing the similarities of the concerns Southern property owners had about the economy with the similar concerns of Northern Republicans. Hayes sought to win acceptance for Republican Party policies as a means to attract Southerners back to national loyalty. He appealed to Confederate veterans as a veteran himself, recognizing Southern loyalty, courage, and resilience, while acknowledging that the war's outcome did not make the Confederate cause wrong. Hayes articulated a vision of equal rights that had first unified the country, and which had been first formulated by Southerners such as Washington, Jefferson, and Madison, coming to pass in the postwar America that would now exist for both whites and blacks. He believed Americans could all reconcile on this principle, which was in harmony with the Constitution. Hayes, although sympathetic to the aspirations of Southern blacks for advancement, security, and rights, took a gradualist approach to race relations, believing that both North and South were equally guilty for the maltreatment of blacks in American society. He believed that equal rights would be assured through an educated population that would raise both races out of poverty and allow blacks to gain the vote on their own merits. Education would ameliorate attitudes and build respect for the law of the land.[5]

Hayes was establishing one-half of the reconciliation process. Recognizing a mutual responsibility for black citizens took the burden off Southerners and raised the issue of equality to a national level responsibility. A shared faith and respect for the Constitution that had existed at the birth of the United States was again possible as a basis for reconciliation. Veterans could rebuild the nation through their display of mutual respect for the common ideals of selfless sacrifices made, patriotic duty performed, and battlefield courage demonstrated. These actions served to promote the redefinition of the North-South relationship to develop shared perspectives and emotions. The spirit of the Revolutionary War, like the Constitution, was a convenient vehicle for rebuilding relations. The American Revolution had united North and South in a common cause for freedom. The Southern cause of independence had overtures of the spirit of 1776, a theme that the Confederacy often promulgated. In 1875, during the observation of the anniversary of the Battle of Bunker Hill, Northerners respectfully accepted this theme as a contingent of military units from Virginia and North Carolina attended the ceremonies to great acclaim.[6]

The devotion of the Southern people to remembering the dead came in time to focus on shared American ideals of liberty rather than on the causes of the war. Through Memorial Day ceremonies held in both the North and South, the memories of the war were able to merge in a common American identity through a shared appreciation of sacrifice. These ceremonies "produced potentially powerful and even spiritually elevated moments in which Americans drew distinct meaning from the past."[7]

The veteran himself became a unifying symbol for the country, reflecting the popular conceptions of heroic citizenship. For the veterans, the memory of combat was a source of personal regeneration and a demonstration of martial heroism. Mutual respect and admiration for personal courage, steadfastness, and obedience to duty became more important to the veterans than the cause and purpose of the war. Significantly, the mutual respect the veterans demonstrated to each other played the most important role in creating a new shared national collective identity. Union veterans increasingly accepted their former foes as noble Americans fighting with courage and devotion for a just constitutional cause—in the same manner as they themselves had for their constitutional cause. There was plenty of glory to share for both sides. Battlefield parks at Chickamauga and Gettysburg were established, monuments dedicated, and reunions held. Robert E. Lee and Ulysses S. Grant became mutual heroes and symbols of reunion.[8]

In 1879, the Richmond Light Infantry Blues participated in the unveiling of the Maj. Gen. George H. Thomas monument in Washington, D.C. Thomas was honored both for his important role as a Union army commander during the war and as a native Virginian. In 1885, four companies of Virginia militia attended the funeral procession of Ulysses S. Grant in New York City. Pallbearers included his wartime comrades Union generals William T. Sherman and Philip H. Sheridan, as well as Confederate generals Joseph E. Johnston and Simon B. Buckner. Flags flew at half-staff throughout Richmond, the former capital of the Confederacy, as the former general-in-chief of the Union armies and president of the United States was laid to rest.[9]

Political equilibrium allowed Southerners to reconcile with their former state adversaries. In 1876, for example, former Republican governor Rufus Bullock returned to Georgia from New York where he had fled in 1870 after secretly resigning. He stood trial on various charges of corruption, conspiracy to defraud the state, and malfeasance. After the prosecution failed to substantiate its case, and two Atlanta juries declared him not guilty, Bullock remained in Atlanta and became one of the city's most prominent citizens. He was president of Atlanta's first cotton mill, president of the English-American Loan Company, senior warden of St. Philip's Episcopal Church, president of the Atlanta Chamber of Commerce, president of the city's Commercial Club, vice president of the Capital City Club, and a member of the high-society Piedmont Driving Club. His significant role in organizing the Atlanta Exposition of 1895 included persuading Booker T. Washington to give a keynote address, one of the most significant speeches marking the New South's presentment of racial accommodation.[10]

One of the most remarkable, and often overlooked, aspects of the political and social equilibrium attained in the South was the appearance and toleration of black militia units. Less than 20 years earlier, the presence or prospect or existence of armed black men created a sense of fear and threat among white Southerners that led to an immediate and often brutally violent reaction. Yet after the Republican Reconstruction governments had been deposed, and black men had been largely removed as the key political actors in the state, the fear disappeared, and there was no need for preventing armed black military units from organizing and drilling in public with weapons. In the New South, black militia units were accepted as part of the fabric of Southern life. Texas in 1882 had 9 black militia companies of 352 men each; Virginia in 1885 had 19 companies of black militiamen of 1,000 men each; North Carolina had 11 black militia companies in 1878 with more than 800 men; and South Carolina had a total of 837 black militiamen in 1891. In 1892, Georgia had 952 black militiamen, and, in 1898, Alabama had a black militia battalion of 181 men. About the same time, Tennessee and Arkansas each maintained 2 black militia companies. Black state militia units in the South represented 20 to 40 percent of the total militia force for each state. Black citizen-soldiers adopted the patriotic mien of white Southerners, and together they openly demonstrated their loyalty to the United States.[11]

Increasingly, the national flag and the Confederate flag could fly together: the stars and stripes held in shared loyalty and esteem, and the stars and bars held in deference and sentimental attachment. The new shared American national identity that emerged was shaped by the large number of Southerners who maintained a cultural conception of the Confederate nation as a component of a Southern collective identity. Popular interest in the Civil War also supported the American national identity. The stories of the war and the great battles overshadowed the political and social issues that led to war and emphasized a shared national pride in the men who fought the war. Especially popular was the series entitled "The Battles and Leaders of the Civil War," a collection of the personal reminiscences of the participants published between 1884 and 1886 in *Century Magazine* that served to "cast a permanent aura of heroism over how Americans thought about war."[12]

The Confederate experience represented both tragedy and triumph, in which Southerners gained regeneration and recognition as Americans. The titanic events of the Civil War required that Southerners derive a larger meaning from the past. The Southern understanding of the war became the means by which reconciliation was possible. Americans recognized that Confederates had fought honorably and with great dedication and courage defending high ideals, the Confederate soldier had demonstrated an admirable martial spirit on the battlefield, Confederate commanders had conducted battles and campaigns with exceptional skill, and the women of the Confederacy had remained unshaken in the face of privation and invasion. By the 1880s, both sides regarded the destruction of slavery as a beneficial and acceptable outcome, and they agreed that blacks were unprepared to exercise the franchise or be trusted with the reins of government, leading to a national acceptance of white racial dominance in the following decade.[13]

Thus, the history of the war accommodated the Southern-Confederate identity and established it as a commensurate aspect of a national American identity. While willing to return to the Union and demonstrating patriotism and love of country as Americans, Southerners also needed to be recognized for having maintained the specific features of honor, patriotism, and duty that defined the Southern-Confederate collective identity. This dual identity came to be accepted as part of reconciliation and allowed Southerners to subscribe to an American national identity but also to demonstrate an attachment to the Southern-Confederate collective identity that served as a unique form of self-definition within the United States. The acceptance of this dual identity allowed the slow and gradual transition from Southerners being cast out as rebels and traitors to being accepted as both Southerners and Americans. The Southern collective memory of the war, invoking shared emotions and consciousness, became an essential part of the dual identity of white Southerners and became an essential part of the reconciliation process. Its enormous power explains why this identity was able to sustain itself until the latter years of the twentieth century.[14]

However, what marked the true reconciliation between North and South was the advent of the New South. The New South was a program touting regional development through a balance of industrial development, diversification of agriculture, and railroad expansion. In essence, the New South's underlying message was that economic development on the model of the North was far more important to the nation than the ephemeral pursuit of social change in the South. The *New York Herald* captured the attitude of many Northerners who recognized the potential benefits of the changed conditions in the Southern states:

> The entire south, so recently antagonized, and exhausting its energies of her people in continuous political strife, is now in a condition of contentment and repose, and as a consequence, we are entitled to look for a steady revival of the important business interests of that section of the Union.[15]

The spirit and optimism of the New South promoters, all of whom were born in the decade before the war, established the political, economic, and cultural relationship that brought about a sense of true reunion through cooperative interaction and interdependence in pursuit of the same lofty goals. They were perceived as more practical and more reasonable than their fathers. Although raised on the arguments that asserted the principles of state sovereignty and justified secession, and fully understanding "what the war was for," the new generation saw no way to advance by attachment to the old ways, but sought inspiration from the present and the future. They favored Northern immigration, believed in railroad expansion and expanding trade and investment, preaching what one historian has described as "laissez faire capitalism, freed of all traditional restraints, together with a new philosophy and a way of life and a new scale of values."[16]

From the perspective of Northerners, Southerners no longer appeared to be the backward outliers who required conversion and renewal to join the

Republican postwar vision of America. Political equilibrium opened their eyes to opportunities for expanding markets and taking advantage of cheap labor, land, and facilities. Untouched and abundant natural resources were available, and railroad development began in earnest in 1879.[17] The New South as myth and reality reshaped the relationship between North and South by creating new political, cultural, and economic relationships.[18] It succeeded in establishing common motivations, actions, and attitudes related to progress and "helped make permanent the very idea of the South."[19]

While Southerners adapted to the new words, ideas, and traditions of Northern capitalism, combining them with extravagant promises of progress and prosperity, they maintained a regional distinctiveness manifested by loyalty to the past demonstrated in a Southern-Confederate collective identity.[20]

Robert L. Taylor, one of the exemplars of the new generation, spoke to an appreciative crowd during his 1886 campaign for governor of Tennessee, presenting a vision of the New South nearly indistinguishable from the vision Charles Sumner, Thaddeus Stevens, or Ben Wade proclaimed as the strategic goals for Reconstruction in 1867. The New South, Taylor proclaimed was built on "the enduring principles of free thought, free action, free labor, [and] free ballot;" the New South, moreover, represented "justice, law, order . . . the education of the masses, the autonomy of the states, constitutional government, one flag and a reunited public." Reconciled and reunited, men of the South and North would, in the bonds of brotherhood, work out "the grand destiny of the new South and the whole nation."[21]

Henry Grady was born in Athens, Georgia, in 1850, the same year as Robert Taylor. His father, a Confederate officer, died in battle in 1864. Grady attended the University of Georgia and became the managing editor of the *Atlanta Constitution*, where he became the advocate of a new South that would rebuild and advance socially and economically through industrial development, political parity, and racial stability. Grady gave a speech to the New England Society of New York in December 1886 that represented the apotheosis of the vision of the New South that allowed the spirit of sectional reconciliation to coalesce into long-term peace. Speaking as a representative Southerner, Grady outlined the South's conversion from sectionalism to national unity and presented a land of unlimited promise with an amenable racial climate, telling his audience that in the New South, "we have sowed towns and cities in the place of theories, and put business above politics." He proclaimed, "The New South is enamored of her new work" with a soul "stirred with the breath of a new life," and "thrilling with the consciousness of growing power and prosperity." Taking the unspoken question head-on, Grady addressed the conditions of blacks in words that would encourage Northern investors with the image of a vast labor pool, while also soothing the moral concerns of his audience. "No section shows a more prosperous laboring population than the Negroes of the South," Grady assured his listeners. Indeed, Southern blacks, he claimed, have "the fullest protection of our laws and the friendship of our people." He admitted that while the law could only give blacks "liberty and enfranchisement," Grady asserted that white Southern "conscience and common sense" would have to do the rest, but he did confide that "if the South holds her reason and integrity," it will continue to keep faith with the black population both now and in the future.[22]

In the person of Henry Grady, the New South represented the means by which the two sections found mutually acceptable values and norms. But the New South spirit of reconciliation only existed as a result of a process of accommodation and compromise that, in turn, came about through the equilibrium brought about by asymmetric warfare. In this reconciliation, which brought about peaceful, cooperative, and trustful relations, the rival sections reunited along common goals, beliefs, symbols, and history that incorporated Southerners as an acceptable group within a common American identity.[23]

Conclusion

When the weaker actor's interests are severely threatened by the dominant actor's efforts to change the status quo, the weaker actor may be willing to take greater risks to protect its interests and initiate asymmetric warfare.[1] Asymmetric warfare is a mode of operation in which a weaker actor uses its strengths and advantages to exploit the weaknesses and vulnerabilities of a dominant actor in order to achieve a level of political equilibrium favorable to the weaker actor. The advantages of a weaker actor are found in the strength of resolve and will and a higher tolerance for absorbing the costs necessary to achieve a favorable outcome.[2]

Asymmetric warfare employs nonviolent means as well as one-sided violence in a number of forms and layers against civilians representing political actors. These means are employed through a combination of different tactics to achieve a destabilizing effect on the dominant actor by challenging the credibility of the dominant actor's military and political commitments to its strategy as well as its willingness to bear the political costs. An equilibrium is achieved when the dominant actor makes concessions and abandons or significantly lowers the strategic goals it was pursuing.

The formal end of the Civil War was marked by individual units of the Confederate army being disarmed and the soldiers being demobilized as parolees, who reverted to the protected status of private citizens. Lee's surrender of the Army of Northern Virginia represented the end of hostilities, but Appomattox was only a symbol; none of the underlying causes of the war had yet been addressed. Although the war appeared to have settled everything, the postwar environment that came about as a result of the Union victory was opaque and filled with potential for future conflict and violence.[3] The most basic process of reconciliation had been delineated in President Lincoln's wartime policies for the restoration of state governments under military occupation. But his policy to deal with the emerging postwar environment and the better peace that would emerge had been outlined only in the broadest terms in his second inaugural address and in his guidance to his commanders in the last weeks of the war.

Extraordinary social, political, and economic challenges confronted the victors. Somehow, these interests had to be addressed in the South through an integrated approach that would rebuild the shattered economy, promote economic growth, and restore democratic institutions, while also establishing a new social fabric that accounted for former slaves, now elevated to the undefined condition of free people. The wartime Freedmen's Bureau was already in place and could have served as the basis for this type of integrated approach, but it was severely undermanned and under resourced. The national shock of the death of Abraham Lincoln at the hands of a pro-Southern assassin only intensified the already existing sentiment and widespread desire to punish Southerners for rebellion, disloyalty, and treason. President Johnson intended to create a political equilibrium by presenting Congress with a fait accompli that would allow the executive and legislative branches to hammer out a consensus vision for postwar America. Unfortunately, the Southern states under Presidential Reconstruction appeared to be returning to the status quo ante bellum, unwilling to demonstrate submission to federal power, placing the freedpeople in a position of subordination to white control and oversight, and sending former Confederates to Congress. The Republicans in Congress rejected this equilibrium and took control of Reconstruction, arguing that they had guided the country through a terrible Civil War, saved the Union, and freed the slaves. As a result, they believed that Congress should determine the course of postwar reconstruction and address the future of the freedmen in the reincarnated American republic. The decisive and total military defeat of the Confederacy seemed to have swept aside any considerations of the complexities of reconstruction and reconciliation.[4]

Congress developed a strategy that used a combination of legal and military force to impose its will on the South with the goal of creating a new equilibrium based on the Republican ideology of freedom, equality, and moral and economic progress. The annihilation of the existing Southern states and the creation of military districts, the reduction of the governments of these former states to a provisional status, and district commanders having the arbitrary power to remove elected officials and replace them with more compliant individuals who could take the ironclad oath were all intended to demonstrate the intent of Congress as the dominant actor to use its power to impose its will on a weaker actor. Federal troops were already in the South, representing the ability of the government to enforce its will and reflecting the expectation of the dominant actor that these forces, although limited in number, would be adequate for military commanders to make a rapid transition from military districts under martial law to readmitted states.

The Congressional Reconstruction strategy intended to re-create the states under a new political order to ensure that the strategic goals of Congress would be accomplished. The strategy was built on the requirement to marginalize white Southern influence in the newly established state governments and fill the power vacuum with loyal whites and free blacks, who now had to be invested as the prime political actors in order to ensure long-term Republican political control. This political base would guide the transformation and integration of these new states into the national postwar economic and social order. As a prerequisite for readmission, constitutional amendments would give the freedmen the political leverage to transform the South through a redefinition of citizenship and political rights. Congress had authority for the enforcement of these amendments as well. Republicans not only had "emancipated the slave and made him a free man," a pro-Republican North Carolina newspaper proclaimed, but they also "armed him with the mighty power of the ballot with which to guard his freedom." This was a powerful, yet flawed assumption of the Congressional strategy. Faith in the power of law to change the postwar conditions in America was not enough. "The mighty power of the ballot" was indeed the means of transformation, but that power could be contested and wrested from the hands into which that power was placed.[5]

Although the Freedmen's Bureau's powers were extended by legislation, it was never adequately resourced or integrated into the Republican Congressional Reconstruction strategy. Federal forces in the South numbered 20,117 men in 1867 but continued to shrink throughout the next 10 years, reaching 6,011 in October of 1876. Commanders in the military departments could shuttle forces from state to state under orders from the War Department, but only on rare occasions were these forces sufficient to do more than providing a token symbolic presence. The limited deployments of small detachments to trouble spots indicated to Southerners that the dominant actor was unwilling to commit the resources necessary to support its strategy and allowed the weaker actor to maintain the initiative.

The Congressional strategy with its goal of consolidating "the American idea of democratic republicanism" to create a true and pure democracy contained a fatal flaw that ensured a renewal of conflict that became asymmetric warfare. Without a process of reconciliation that provided some level of mutual engagement to secure accommodations between the former antagonists, the Republican Congressional strategic goals left white Southerners outside of the political mainstream. Quite simply, they were not included in the Republican considerations for the future of the nation. Because the Republicans feared the power of a restored national Democratic party to undo the hard-won fruits of victory, differences could not be resolved through normal political means.[6]

This marginalization presented a direct threat to white Southerners, especially small farmers and the working class, who faced the unacceptable prospect of the loss of both power and liberty. The Congressional Reconstruction strategy represented an external threat supported by military occupation and martial law. The only choices for white Southerners were passive accommodation and surrender, or resistance. No longer possessing the political means to establish a viable separate and independent government, and unwilling to accept the new conditions imposed on them, white Southerners in each state turned to challenging the Republican's authority and legitimacy through asymmetric means, avoiding the strengths and exploiting the weaknesses of the state governments.

These conditions represented a combination of complex social phenomena that pushed white Southerners back toward a Confederate wartime national identity that formed the basis of a Southern-Confederate collective identity. This identity was based on particular perceptions, beliefs, attitudes, and motivations that provided a need for unity in the face of the threat to the survival of the collective identity. This uniting of various groups through internal social links to address real and perceived grievances and injustices and to protect common interests defined by the collective identity based on a selective interpretation of events led to mobilization—a threshold for a decision for action. Mobilization represented the transition from a collective disposition into a collective action that became the basis for the initiation of asymmetric warfare against the dominant actor. The process of collective identity mobilization determined who would form the resistance and who cooperated with whom and marked an initial departure point for action. Localities provided the incentives for individuals to carry out violence, and as violent acts became more effective, the threshold of support was crossed, bringing a higher level of incentive to take action to contribute to the collective identity's defense.[7]

Asymmetric warfare emerged because of the relative strengths of the antagonists. Clearly, the postwar South was unable to muster a collective regional resistance. However, in the face of a powerful dominant actor that already had troops on occupation duty and could conceivably draw on additional military manpower as needed, Southerners could organize a low-level resistance at the local level—in counties and parishes across the South. The use of violence was intended to impose costs on the dominant actor while remaining unaffected by the counter actions of either the state governments or federal authorities. Success in asymmetric warfare requires that the weaker actor demonstrate both the durability of the mobilized collective identity and an intensity of interest and will in the face of the dominant actor's superior power and capability to achieve decisive results. This is accomplished by the weaker actor displaying a higher level of motivation, unity, determination, emotional involvement, and commitment to its goals than the dominant actor, while also demonstrating the willingness to use unorthodox methods and go to extreme limits to accomplish those goals. Durability and intensity assist in consolidating the support to the resistance and strengthening support to encourage more individuals to cross the threshold from passive to active participation.[8]

The battlefield of the South's asymmetric warfare effort was both political and psychological. This battlefield had two objectives: first, the state governments could not be given the breathing space to establish their authority and legitimacy; second, the resistance against corrupt state governments run by alien Northern interlopers, Southern political turncoats, and a mass of degraded and unfit freedmen would have to became a dominant image in the minds of the population of the North. As the weaker actor, the physical violence in the conduct of asymmetric warfare would be episodic and protracted. The nature of the violence could be both intentional, discriminate, targeted acts of violence coexisting with indiscriminate, random, and criminal violence that layered over each other, creating an environment of disorder and fear that served the goals of the resistance. This layered violence was the means by which asymmetric warfare would achieve its goals. Violence was employed in various forms by various groups for various purposes against specific targets to force compliance. Selective violence targeted specific individuals for political purposes. Indiscriminate violence against certain collective or individual targets was intended to send a message, demonstrate the resistance's power and control, and highlight the weakness of the state authorities to provide protection. Violent criminal activity, especially murder and arson, although not usually related to the resistance, often served the interests of the resistance.

The dilemma presented to the dominant actor was that the layered violence directed against the freedmen and other political actors, who held the key to political power and legitimacy of the Republican state governments, could not be controlled without a massive (and politically unacceptable) escalation of force. Outmatched on this battlefield, the dominant actor made several attempts to seize the initiative through a combination of legal and military means but did not have the will to bear the costs involved, nor could the dominant actor match the weaker actor's level of intensity of commitment indicated by armed paramilitary groups signaling a willingness to escalate the current conflict into a bloody race war of annihilation.

With the implementation of the Congressional Reconstruction strategy, white Southerners responded by passing through the first phase of asymmetric warfare. The transition from Presidential to Congressional Reconstruction indicated that the South would be placed in a subordinate position and, because of that power imbalance, the sociopolitical goals of the dominant actor that would be imposed were completely contradictory to the weaker actor's own interests and goals. The transition from military districts to readmitted states under Republican control initiated the second phase of asymmetric warfare.

The Southern-Confederate collective identity coalesced and manifested itself as an active violent resistance to reassert and regain power in the form of the Ku-Klux phenomenon that swept the South between 1867 and 1870. Neither centrally organized nor directed, and neither a guerrilla nor a terrorist movement, the Ku-Klux phenomenon was instead a reflection of identity and defiance that employed layered violence along with the threat of violence to serve the purpose to weaken and delegitimize the Republican Reconstruction state governments. The unorthodox and completely unexpected rise of costumed men and horses as agents of a mobilized collective identity was a decisive act that changed the political strategic landscape of the Reconstruction South. Surprise, based on improbable actions and responses, is an essential element of asymmetric warfare. The Ku-Klux, operating as a mysterious, all-pervasive entity that suddenly appeared to be legion in the South, was devastatingly effective both as a factor in changing the threshold of support in favor of white Southerners and in demonstrating the inability of the Republican state governments to maintain security or stake any claim to legitimacy.

Asymmetric warfare relies on layers of violence operating in a milieu of disorder and insecurity for those who are the targets of the violence. These layers of violence are the operational instrument of asymmetric warfare. Because the violence had no central direction or unified purpose, it was, at times, highly politically focused and executed with military precision; at other times, random and opportunistic; and at still other times, capricious and irrational, with each and every act of violence serving the overall interest of the asymmetric actor. Thus, Ku-Klux violence was an effective means to consolidate control as well as to silence or eliminate high-profile political actors. Indiscriminate violence directed against freedmen usually had limited utility, but the threat of such violence did gain results.

The federal government struggled to assist in stabilizing the Republican state governments that, under the Congressional Reconstruction strategy, had a semblance of sovereign power but represented the state in name only. Establishing political legitimacy through effective institutions and popular support for these governments could not be imposed from the outside, and the governments themselves could not respond effectively to the asymmetric threat of the Southern-Confederate mobilized collective identity.

As asymmetric actors, the Ku-Klux succeeded in seizing and holding the imagination of the nation in such a decisive manner that the dominant actor never was able to recover the initiative. In this phase, the weaker actor also gained the dominance of the information battlefield and forced a number of state Republican Reconstruction governments to take military and legal action that ironically contributed to their downfall. The state governors faced catastrophic consequences from a powerful organized resistance if they attempted to pursue suppression measures too aggressively. Even the federal reaction with the declaration of martial law and the use of U.S. military and federal law enforcement in South Carolina to suppress the Ku-Klux was far too little and far too late to regain the initiative. Instead, it indicated the crumbling will of the dominant actor to pursue its strategic goals and an unwillingness to bear the costs of protecting the Republican Reconstruction governments.

As Republican governments were replaced in one state after another in the election cycles, the states that remained in Republican hands faced the culminating action of the third phase of asymmetric warfare—the bulldozers. This phase was uniquely demonstrated in Louisiana and South Carolina, where white Southerners organized heavily armed paramilitary units operating in conjunction with Democratic gubernatorial candidates as a means of demonstrating an overmatching capability against the Republican governments to challenge them. The bulldozers largely avoided using violence but, instead, relied on psychological dominance, using open displays of force to intimidate voters and disrupt political meetings. As the dominant actor's will to pursue its strategic goals continued to drain away, the appeals for assistance from the state governors went largely unheeded, and even when federal troops were dispatched as a result of an urgent request, the paramilitary units simply disappeared. However, it was well understood that these units could reassemble and be employed at any time and in any location in the state. In Louisiana, the White League paramilitary units were so well organized and led that while operating as the military arm of the state Democrats, they easily took control of the state capital, installed a shadow government, and defied both the state and federal authorities to oppose them. The state and national elections in 1876, fraught as they were with controversy in the Southern states and distinguished by political turmoil at the national level, actually initiated the fourth phase of asymmetric warfare—political equilibrium. Public opinion had turned against the Republican strategic goals, the Congress no longer had the political will to sustain the strategy, and the new president-elect had already signaled his willingness to reach some accommodation with white Southern leaders controlling the state governments. When the dominant actor was no longer politically capable of achieving its original strategic goals, it had to accept far more modest but politically achievable goals. White Southerners had achieved their minimum goals as well, and this accommodation established a consensual framework that allowed the two actors to move toward reconciliation. A new consensus emerged, allowing a Southern narrative representing its truth about the war to be articulated along with an acceptance of a Southern-Confederate collective identity merged within an American national identity.

As Henry Grady and other acolytes of the New South professed, the South did offer great promise in the years following political equilibrium and reconciliation. The South did indeed enjoy a prosperity marked by rising wages, growing cities, and diversified industrial growth in coal, steel, timber, cotton textiles, and tobacco, but there was something incomplete and hollow about this New South that emerged. In actuality, it was more *different* than new—ahead were internal political and social upheavals, institutionalized segregation and disfranchisement, a cultural renaissance, and a struggle that consumed three generations of white Southerners to hold on to what had been gained through nearly a decade of asymmetric warfare.[9]

Notes

INTRODUCTION

1. David Madden, "Afterword: Untangling the Webs of the Civil War and Reconstruction in the Popular Culture Imagination," in *The Civil War in Popular Culture: Memory and Meaning*, eds. Lawrence J. Kreiser and Randal Allred (Lexington: University Press of Kentucky, 2014), p. 224.

2. Revisionists who stressed that more military force was the answer to the problems of the South most likely did not have the same enthusiasm for using military force to address similar conditions in Vietnam, Iraq, and Afghanistan.

3. Laura F. Edwards, "Reconstruction and the History of Governance," in *The World the Civil War Made*, eds. Gregory P. Downs and Kate Masur (Chapel Hill: University of North Carolina Press, 2015), pp. 22–26.

4. Eric Foner, "Reconstruction Revisited," *American History* 10, no. 4 (December 1982): p. 82.

5. Ibid., p. 95. The revitalization of W. E. B. DuBois's book *Black Reconstruction* is instructive. DuBois presented two themes that are now very much a part of the current direction of Reconstruction studies. The first is the erroneous, but emotionally satisfying, idea that the majority of Americans in the North believed in black equality; the second is the toxic, but equally emotionally satisfying, idea that Southern whites were nothing more than a deranged population consumed with "deliberate vengeance and hate." Much of the dominant political dialogue today in the United States is shaped by this narrative. See Sharon D. Kennedy-Nolle, *Writing Reconstruction: Race, Gender, and Citizenship in the Postwar South* (Chapel Hill: University of North Carolina Press, 2015), p. 295.

6. Madden, "Afterword," p. 224.

7. Douglas R. Egerton, *The Wars of Reconstruction: The Brief, Violent History of America's Most Progressive Era* (New York: Bloomsbury Press, 2014), p. 21. See also F. Michael Higginbotham, "Maintaining White Dominance during Reconstruction," in *Ghosts of Jim Crow* (New York: New York University Press, 2013), pp. 63–84.

8. Graeme Simpson, "Reconstruction and Reconciliation: Emerging from Transition," *Development in Practice* 7, no. 4 (November 1997): p. 477.

9. Michael Vlahos, *Fighting Identity: Sacred War and World Change* (Westport, CT: Praeger Security International, 2009), pp. 76–77. Vlahos describes war as a "ritual of identity."

CHAPTER 1

1. David L. Buffaloe, "Defining Asymmetric Warfare," The Land Warfare Papers, no. 58 (Arlington, VA: The Institute of Land Warfare, Association of the United States Army, 2006), p. 3. Buffaloe reviews the difficulty the U.S. military had in applying the term and concludes that "due to a lack of concrete understanding, the term became meaningless." Quote from Gen. Montgomery Meigs, p. 11. Patricia L. Sullivan, "War Aims and War Outcomes: Why Powerful States Lose Limited Wars," *Journal of Conflict Resolution* 51, no. 3 (June 2007): p. 509.

2. Raymond W. Mack and Richard C. Snyder, "The Analysis of Social Conflict—Toward an Overview and Synthesis," *Conflict Resolution* 1, no. 2 (June 1957): pp. 217–220, passim. Josef Schroefl and Stuart J. Kaufman, "Hybrid Actors, Tactical Variety: Rethinking Asymmetric and Hybrid War," *Studies in Conflict and Terrorism* 37, no. 10 (2014): p. 865.

3. J. G. Eaton, "The Beauty of Asymmetry: An Examination of the Context and Practice of Asymmetric and Unconventional Warfare from a Western/Centrist Perspective," *Defence Studies* 2, no. 1 (Spring 2002): pp. 53, 67.

4. Michael G. Findley and Scott Edwards, "Accounting for the Unaccounted: Weak-Actor Structure in Asymmetric Wars," *International Studies Quarterly* 51, no. 3 (September 2007): pp. 583–584.

5. Stathis N. Kalyvas, "Warfare in Civil Wars," in *Rethinking the Nature of War*, eds. Isabelle Duyvesteyn and Jan Angstrom (New York: Frank Cass, 2005), pp. 88–91, passim.

6. Clinton J. Ancker and Michael D. Burke, "Doctrine for Asymmetric Warfare," *Military Review* 83, no. 4 (July–August 2003): p. 18.

7. Kalyvas, "Warfare in Civil Wars," p. 96.

8. Stathis N. Kalyvas, "The Ontology of Political Violence: Actions and Identities in Civil Wars," *Perspectives on Politics* 1, no. 3, (September 2003): p. 480.

9. S. J. Ball-Rokeach, "Normative and Deviant Violence from a Conflict Perspective," *Social Problems* 28, no. 1 (October 1980): p. 47. Ball-Rokeach defines the two terms: "Violence from 'normal' social and personal processes may be called *social* violence to distinguish it from 'asocial' violence caused by 'abnormal' or deficit states." In the voluminous testimonies given regarding violence in the Southern states during Reconstruction, these terms work well to ascribe accounts of whippings and beatings as examples of social violence, and murder, rape, and assassination as examples of asocial violence. The perpetrators of violence in Phase II of asymmetric warfare in the South may not have perceived a distinction, depending on their particular perception of the utility of the violence.

10. Kalyvas, "The Ontology of Political Violence," pp. 230–232, 234, 242, 244. C. R. Mitchell, "Evaluating Conflict," *Journal of Peace Research* 17, no. 1 (1980): p. 65.

11. Ekaterina Stepanova, *Terrorism in Asymmetrical Conflict: Ideological and Structural Aspects*, SIPRI Research report no. 23 (New York: Oxford University

Press, 2008), pp. 12–13, 25. The violence employed in asymmetric warfare is not terrorism—"a performance that involves the use or threat to use violence against civilians, but which is staged specifically for someone else to watch" (p. 13). Although layered violence in asymmetric warfare certainly can involve terrorist tactics, "the fact that a group uses terrorist means in the name of a political goal does not necessarily delegitimize the goal itself" (p. 12). In the nineteenth century, terrorist means were used by radical left-wing ideologies, and, strictly speaking, violence in the South during the Reconstruction period would not have been considered terrorism.

12. Mitchell, "Evaluating Conflict," p. 66.

13. Michael L. Gross, "Asymmetric War, Symmetrical Intentions: Killing Civilians in Modern Armed Conflict," *Global Crime* 10, no. 4 (November 2009): pp. 322, 324–325, 328.

14. Stathis N. Kalyvas, *The Logic of Violence in Civil War* (New York: Cambridge University Press, 2008), pp. 143, 145, 364.

15. Ibid., pp. 20, 25–27, 31, 38, 147.

16. Thomas C. Davis, "Revisiting Group Attachment: Ethnic and National Identity," *Political Psychology* 20, no. 1 (March 1999): pp. 25–26, 31.

17. Kay Deaux and Tracy McLaughlin-Volpe, "An Organizing Framework for Collective Identity: Articulation and Significance of Multidimensionality," *Psychological Bulletin* 130, no. 1 (2004): pp. 82–83, 96.

18. Errol A. Henderson, "Culture or Contiguity: Ethnic Conflict, the Similarity of States, and the Onset of War, 1820–1989," *Journal of Conflict Resolution* 41, no. 5 (October 1997): p. 657.

19. Kenneth D. Bush and E. Fuat Keyman "Identity-Based Conflict: Rethinking Security in a Post-Cold War World," *Global Governance* 3, no. 3 (September–December 1997): pp. 312–315, 317–319, 325–326.

20. Christine Flesher Fominaya, "Collective Identity in Social Movements: Central Concepts and Debates," *Sociology Compass* 4, no. 6 (2010): pp. 394–395, 397–398.

21. Kalyvas, "The Ontology of Political Violence," pp. 475, 479, 486–487.

22. E. J. R. Chamberlain, "Asymmetry: What Is It and What Does It Mean for the British Armed Forces?" *Defence Studies* 3, no. 1 (Spring 2003): p. 20.

23. Andrew J. Williams, "Reconstruction before the Marshall Plan," *Review of International Studies* 31, no. 3 (July 2005): p. 542. Williams observes that reconstruction in the nineteenth century had a "positive economic normative connotation."

24. Sullivan, "War Aims and War Outcomes," pp. 499–501. Sullivan describes cost tolerance as the extent to which an actor is willing to absorb the human and material costs to achieve its objectives. See also Philip Marshall Brown, "Postwar Reconstruction," *World Affairs* 170, no. 1 (Summer 2007): pp. 33–35. Roger MacGinty, "The Pre-War Reconstruction of Post-War Iraq," *Third World Quarterly* 24, no. 4 (August 2003): pp. 603–604.

25. David E. Cunningham, Kristin Skrede Gleditsch, and Idean Salehyan, "It Takes Two: A Dyadic Analysis of Civil War Duration and Outcome," *Journal of Conflict Resolution* 53, no. 4 (August 2009): p. 575.

26. Helmut Dubiel, "Cultivated Conflicts," *Political Theory* 26, no. 2 (April 1998): pp. 217–218.

27. Findley and Edwards, "Weak-Actor Structure in Asymmetric Wars," pp. 587–588, 591.

CHAPTER 2

1. Kent T. Dollar, Larry H. Whittaker, W. Calvin Dickinson, eds., *Sister States, Enemy States: The Civil War in Kentucky and Tennessee* (Lexington: University Press of Kentucky, 2009), pp. 300–303, 313.

2. William C. Harris, *With Charity for All: Lincoln and the Restoration of the Union* (Lexington: University Press of Kentucky, 1997), pp. 9, 130–134.

3. Isaac N. Arnold, *The Life of Abraham Lincoln* (Chicago: A. C. McClung, 1909), pp. 421–422. Sherman's recollection was included in a letter he sent to Arnold in 1872.

4. Charles Royster, ed., *Memoirs of General W. T. Sherman* (New York: Viking Press, 1990), pp. 314–317. See also Harris, *With Charity for All*, pp. 248–249.

5. J. A. Campbell to Maj. Gen. Godfery Weitzel, April 7, 1865. The War Department, *The War of Rebellion: A Compilation of the Official Records of the Union and Confederate Armies*, Series 1, Vol. 46, part 3 (Washington, D.C.: U.S. Government Printing Office, 1895), p. 657.

6. Clarence Buel and Robert U. Johnson, eds., *The Way to Appomattox: Battles and Leaders of the Civil War Vol IV* (New York: Castle Books, 1956), p. 728.

7. John G. Sproat, "Blueprint for Radical Reconstruction," *Journal of Southern History* 23, no. 1 (February 1957): pp. 34, 36.

8. Ibid., pp. 40, 44.

9. *The Old North State* (Salisbury, NC), October 6, 1866.

10. Jacob Oser, *Henry George* (New York: Twayne Publishers, 1974), p. 24.

11. Assistant Secretary of War John A. Campbell used these prescient and perceptive words in his message to Maj. Gen. Weitzel in April of 1865. See fn 5.

12. Kenneth M. Stampp, *The Era of Reconstruction 1865–1877* (New York: Vintage Books, 1965), pp. 54–56, 59.

13. An article in the *Western Democrat* (Charlotte, NC), July 25, 1865, reflected a common frustration that certainly influenced the creation of the Black Codes. "The Negro must be made to work with promptness and regularity. If everybody North and South stopped talking about Negro equality and Negro suffrage and sought, instead to inaugurate some fair measure to compel him to work, they will do more good for the Negro in that way than any other." An opposing view on the Black Codes was offered in the *Sunbury American* (Sunbury, PA), September 12, 1868.

14. *Anderson Intelligencer* (Anderson Court House, SC), December 13, 1866.

15. Egerton, *Wars of Reconstruction*, p. 173.

16. George P. Smith, "Republican Reconstruction and Section Two of the Fourteenth Amendment," *Western Political Quarterly* 23, no. 4 (December 1970): pp. 832–833, 838. K. Stephen Prince, *Stories of the South: Race and the Reconstruction of Southern Identity, 1865–1915* (Chapel Hill: University of North Carolina Press, 2014), pp. 19, 32–33, 39.

17. Ironically, in that same month, President Johnson had signed the Southern Homestead Act, opening 46 million acres of public land in Alabama, Mississippi,

Arkansas, Florida, and Louisiana to any settler regardless of color. The prospect of free land to freedmen could have had the transformative effect the radical faction desired if this opportunity had been incorporated into their strategy.

18. Egerton, *Wars of Reconstruction*, p. 105.

19. U.S. Congress, Joint Committee on Reconstruction, *Report of the Joint Committee on Reconstruction*, 39th Cong., 1st sess., 1865–1866, pp. x–xi, xix.

20. Ibid., pp. x, xxi.

21. Ibid., pp. xvii–xviii.

22. Ibid., pp. xxi, xxiii. Alexander H. Stephens, former vice president of the Confederacy, a personal friend of Abraham Lincoln, and at the time waiting to be seated as a senator from Georgia, testified before the Reconstruction committee and made this statement: "Universal suffrage among the Negroes, as they now are, would be regarded as about as great a political evil as could befall the people of Georgia." While it is easy to dismiss this statement now, as it was by the committee then, such an honest and forthright answer should have signaled danger. *The Yorkville Enquirer* (SC), April 26, 1866. See also a summary and an assessment of the report in the *New South* (Beaufort, SC), June 15, 1866.

23. A Northern correspondent spoke to a Virginia planter in June of 1865 about the mood of Southerners in the wake of defeat. The response to his question was telling. The planter believed that while the people of the South were "subjugated, conquered, and . . . must submit to whatever may be inflicted upon them," he also added that "they are not going to endure the infamous charge of having committed treason." Quoted in Stephen Budiansky, *The Bloody Shirt: Terror after Appomattox* (New York: Viking Penguin, 2008), p. 34.

24. Ibid., p. x.

25. Bill Kissane, ed., *After Civil War: Diversion, Reconstruction and Reconciliation in Contemporary Europe* (Philadelphia: University of Pennsylvania Press, 2014), pp. 274, 277, 282, 280.

CHAPTER 3

1. Andrew L. Slap, *The Doom of Reconstruction: Liberal Republicans in the Civil War Era* (New York: Fordham University Press, 2006), p. 84.

2. George P. Smith, "Republican Reconstruction and Section Two of the Fourteenth Amendment," *Western Political Quarterly* 23, no. 1 (December 1970): pp. 832–833, 838, 840, 843, 852–853, 858. Albert Bergesen, "Nation-Building and Constitutional Amendments: The Role of the Thirteenth, Fourteenth, and Fifteenth Amendments in the Legal Reconstitution of the American Polity Following the Civil War," *Pacific Sociological Review* 24, no. 1 (January 1981): pp. 4–5, 7, 13. Slap, *Doom of Reconstruction*, pp. 85–86. Slap notes that liberal Republicans were against limitations on former Confederates, advocating that blacks and whites in the South should have equal access to the vote. This became one of the major reasons for the liberals to break away from the Grant administration in 1872 and advocate compromise and conciliation in 1876.

3. Quoted in the *Old North State* (Salisbury, NC), October 20, 1866. Prince, *Stories of the South*, pp. 25–26.

4. Susannah J. Vral, ed. *Civil War Citizens: Race, Ethnicity, and Identity in America's Bloodiest Conflict* (New York: New York University Press, 2016), p. 4. Vral notes that citizenship was considered "the prize of the age." Until the Civil Rights Act of 1866, and the passage of the Fourteenth (1868) and Fifteenth (1870) Amendments, state legislatures controlled citizenship and determined voting qualifications. Northern states had numerous restrictions on black voters and limited their access to courts and schools. Only Vermont (which had a total of 80 blacks in the state) and New Hampshire (which had 190 blacks in the state) allowed blacks to vote without a property qualification. See Joseph P. Reidy, "The African American Struggle for Citizenship Rights in the Northern States During the Civil War," in Vral, ed., *Civil War Citizens*, pp. 215, 217–218. See also the *Daily Phoenix* (Columbia, SC), June 24, 1866, for a summary of the amendment, and the *Evening Telegraph* (Philadelphia, PA), September 3, 1866, for commentary.

5. Act of April 9, 1866 (Civil Rights Act), Public Law 39–26, 14 STAT 27, National Archives and Records Administration, National Archives Identifier 299820; Series: Enrolled Acts and Resolutions of Congress, 1789–2011, Record Group 11: General Records of the United States Government, 1778–2006.

6. Egerton, *Wars of Reconstruction*, pp. 97, 105.

7. *The Wheeling Intelligencer* (WV), June 2, 1866.

8. William Gillette, *Retreat from Reconstruction 1869–1879* (Baton Rouge: Louisiana State University Press, 1982), p. 6.

9. *The Southerner* (Tarboro, NC), June 30, 1866.

10. *Knoxville Whig* (TN), March 14, 1866.

11. Reported in *Public Ledger* (Memphis, TN), May 1, 1866.

12. Jeff Spinner-Halev and Elizabeth Theiss-Morse, "National Identity and Self-Esteem," *Perspective on Politics* 1, no. 3 (September 2003): pp. 517–519.

13. *Public Ledger* (Memphis, TN), April 28, 1866.

14. *The Western Democrat* (Charlotte, NC), May 8, 1866.

15. *Anderson Intelligencer* (Anderson Court House, SC), May 31, 1866.

16. *The Old North State* (Salisbury, NC), September 29, 1866.

17. *The Daily Clarion* (Jackson, MS), December 19, 1866. *Charleston Daily News* (SC), September 1, 1866.

18. *Staunton Spectator and General Advertiser* (VA), December, 4 1866.

19. The *Philadelphia Inquirer* (PA) reflected Senator Sumner's position. The Confederate states by violence and rebellion set aside and destroyed the constitutional government of the state—Congress finds no government organized or officers qualified in accordance with the Constitution. "The Government set up by the President was provisional and needed the approval of Congress. Consequently, all that followed was immaterial and of no avail." No legal government exists in the Southern states. Those governments, therefore, exist "upon sufferance, and may be abolished whenever Congress chooses to exercise its authority." Reprinted in the *Weekly Standard* (Raleigh, NC), December 20, 1866. Concern and reaction is found in the *Nashville Union and Dispatch* (TN), December 6, 1866, and reprinted in the *Edgefield Advertiser* (Edgefield, SC), December 12, 1866. "If the people stand for such an outrage, then indeed will they prove fit subjects for the vilest despotism the world has ever known." See also the *Dallas Herald* (TX), December 29, 1866.

20. See J. Morgan Kousser, "Post-Reconstruction Suffrage Restrictions in Tennessee: A New Look at the V. O. Key Thesis," *Political Science Quarterly* 88, no. 4 (December 1973): pp. 659, 661, 672.

21. William A. Russ, Jr., "Registration and Disfranchisement under Radical Reconstruction," *Mississippi Valley Historical Review* 21, no. 2 (September 1934): pp. 166–167, 169–171. See also Russ, "The Negro and White Disfranchisement during Radical Reconstruction," *Journal of Negro History* 19, no. 2 (April 1934): p. 187.

22. Yehudith Auerbach, "The Reconciliation Pyramid: A Narrative-Based Framework for Analyzing Identity Conflicts," *Political Psychology* 30, no. 2 (April 2009): p. 294. Richard K. Herrmann, Pierangelo Isernia, and Paolo Segatti, "Attachment to the Nation and International Relations: Dimensions of Identity and their Relationship to War and Peace," *Political Psychology* 30, no. 5 (October 2009): pp. 723–725.

23. *Western Reserve Chronicle* (Warren, OH), July 3, 1867.

24. *Delaware Gazette* (OH), August 2, 1867.

25. *The Daily Evening Telegraph* (Philadelphia, PA), October 24, 1867.

26. William A. Dunning, "Military Government in the South During Reconstruction," *Political Science Quarterly* 12, no. 3 (September 1897): pp. 385–387, 389. The Dunning School, as it is known, has for the most part been abjured by historians. Despite an approach that illustrates a clear bias, Dunning certainly still has useful insights worthy of consideration for those who search diligently. Nevertheless, anyone who quotes or refers to Dunning outside of scholarly discourse does so at his peril. See, for example, Ta-Nehisi Coates's attack on Hillary Clinton for her oblique reference to Dunning's view of Reconstruction that was not in conformity with the standard accepted narrative. Ta-Nehisi Coates, "Hillary Clinton Goes Back to the Dunning School," *Atlantic*, January 26, 2016, http://www.theatlantic .com/politics/archive/2016/01/hillary-clinton-reconstruction/427095.

27. District One: Virginia under Maj. Gen. John M. Schofield; District Two: North and South Carolina under Maj. Gen. Daniel E. Sickles; District Three: Georgia, Florida, and Alabama under Maj. Gen. John Pope; District Four: Mississippi and Arkansas under Maj. Gen. E. O. C. Ord; District Five: Louisiana and Texas under Maj. Gen. Philip Sheridan.

28. Marvin Fletcher, "The Negro Volunteer in Reconstruction, 1865–66," *Military Affairs* 32, no. 3 (December 1968): p. 130. The presence of black troops had been a problem since the war, often leading to confrontations and riots that quickly turned into wholesale attacks on black neighborhoods. This happened in the North a number of times in 1863. In Harrisburg, Pennsylvania, Union troops rampaged through the black housing after an altercation in a bar over payment. The draft riot in New York City in 1863 saw massive destruction of black property and severe beatings of blacks on the streets. In Wilmington, North Carolina, in July 1865, white and black units were engaged in a riot, as were black and white troops in Washington, D.C., in August of 1866. Discharged blacks were also a source of confrontation. In Butler, Pennsylvania, in May of 1866, 60–70 discharged soldiers were arrested for drunk and disorderly conduct. Shots were fired at police, which led to a full-scale riot with white mobs destroying black houses. So, the riots in Memphis, Mobile, and New Orleans were not out of the ordinary. They tended

to start from the same source, regardless of location, and they involved black soldiers or discharged soldiers. In nearly every instance, rioters attacked black neighborhoods. See the *Staunton Spectator and General Advertiser* (VA), August, 14, 1866; *Weekly North-Carolina Standard* (Raleigh, NC), May 22, 1867; *American Citizen* (Butler, PA), May 23, 1866; *The Smoky Hill and Republican Union* (Junction City, KS), July 25, 1863; *Alexandria Gazette* (VA), May 24, 1863, and July 14–15, 1863; *The Western Democrat* (Charlotte, NC), July 25, 1865; *Democrat and Sentinel* (Ellensburg, PA), August 9, 1866; *The Evansville Journal* (IL), May 4, 1866.

29. Mark V. Wetherington, *Plain Folk's Fight: The Civil War and Reconstruction in Piney Woods Georgia* (Chapel Hill: University of North Carolina Press, 2005), pp. 221–227.

30. Stathis N. Kalyvas, *Logic of Violence in Civil War*, pp. 83–84.

31. Michael W. Fitzgerald, "Radical Republicanism and the White Yeomanry during Alabama Reconstruction, 1865–1868," *Journal of Southern History* 54, no. 4 (November 1988): pp. 569, 571, 578. Wetherington, *Plain Folk's Fight*, pp. 213–225.

32. James E. Sefton, *The United States Army and Reconstruction 1865–1877* (Baton Rouge: Louisiana State University Press, 1967), pp. 25–31, 37–38. See also William L. Richter, "'Shoot or Get Out of the Way!': The Murder of Texas Freedmen's Bureau Agent William G. Kirkman by Cullen Baker—and the Historians," in *Still the Arena of Civil War: Violence and Turmoil in Reconstruction Texas, 1865–1874*, ed. Kenneth W. Howell (Denton, TX: University of North Texas Press, 2012), pp. 66–67.

33. Thomas Edward Flores and Ifran Nooruddin, "The Effect of Elections on Postconflict Peace and Reconstruction," *Journal of Politics* 74, no. 2 (April 2002): pp. 558–561, passim. See also Gojko Vuckovic, "Promoting Peace and Democracy in the Aftermath of the Balkan Wars: Comparative Assessment of the Democratization and Institution-Building Processes in Croatia, Bosnia and Herzegovina and Former Yugoslavia," *World Affairs* 162, no. 1 (Summer 1999): pp. 4–5, 10.

34. Simpson, "Reconstruction and Reconciliation," pp. 475–478. See also General Orders 101 from the Headquarters Second Military District dated October 18, 1867, prohibiting "violence or threats of violence against anyone registering or exercising the right to vote." For any such violations reported to military authorities, the commander had arrest and trial authority. Printed in the *Old North State* (Salisbury, NC), November 12, 1867.

35. Printed in the *Tri-Weekly Standard* (Raleigh, NC), August 24, 1867, and also printed in Brownlow's highly partisan *Knoxville Whig* (TN), September 4, 1867, which emphasized Pope's belief that the removal of dissident elements—all former Confederates—from the South would rapidly restore peace and security.

36. Eric Foner, *Reconstruction: America's Unfinished Revolution, 1863–1877* (New York: Harper and Row, 1988), p. 324. Russ, "Registration and Disfranchisement under Radical Reconstruction," pp. 166–167, 169–171.

37. Albert Bergesen, "Nation-Building and Constitutional Amendments: The Role of the Thirteenth, Fourteenth, and Fifteenth Amendments in the Legal Reconstitution of the American Polity Following the Civil War," *Pacific Sociological Review* 24, no. 1 (January 1981): pp. 9, 12–13.

38. Flores and Nooruddin, "The Effect of Elections on Postconflict Peace and Reconstruction," pp. 558–561, 563, 568. Otto H. Olsen, "North Carolina: An

Incongruous Presence," in *Reconstruction and Redemption in the South*, ed. Otto Olsen (Baton Rouge: Louisiana State University Press, 1980), pp. 166–167. A North Carolina newspaper described the categories of potential white voters in the state under the Reconstruction Act. The Radicals were the only organized party in the state, described as "a set of desperate politicians, ever on the lookout for the popular current on which they expect to float into office, while professing to feel the greatest possible affection for the Negro in the hope of securing his vote to help them in office." Other types of politicians were identified—the staunch Union men who held to their convictions throughout the war; those who had defended the constitutional right to secession, "until convinced of their error by the success of the federal armies"; and the miscreants—rabid secessionists before and during the war until it became clear that the war was lost, "then swore they had been Union men all the time." *The Old North State* (Salisbury, NC), October 3, 1867.

39. From the *Raleigh Sentinel* (SC), November 1, 1867, as reported in the *Old North State* (Salisbury, NC), November 12, 1867. Charles Robson, "Col. D. M. Carter North Carolina," *Representative Men of the South* (Philadelphia: Charles Robson, 1880), pp. 418–420, http://archive.org/stream/representativeme00robs#page/418/mode/2up.

40. *Wilmington Journal* (NC), November 1, 1866.

41. *The Lancaster Ledger* (SC), November 21, 1867.

42. *The Daily Phoenix* (Columbia, SC), November 8, 1867.

43. *Richmond Whig* quoted in *The Old North State* (Salisbury, NC), reprinted November 12, 1867.

44. John Kekes, "War," *Philosophy* 85, no. 332 (April 2010): pp. 216–218.

45. Jan E. Stets and Peter J. Burke, "Identity Theory and Social Identity Theory," *Social Psychology Quarterly* 63, no. 3 (September 2000): pp. 281, 283.

CHAPTER 4

1. Office of the President, *Messages from the President of the United States to the Two Houses of Congress at the Commencement of the Third Session of the Fortieth Congress with the Reports of the Heads of the Departments and Selections from Accompanying Documents* (Washington, D.C.: Government Printing Office, 1869), p. 1.

2. William A. Russ, Jr., "Was There Danger of a Second Civil War During Reconstruction?" *Mississippi Valley Historical Review* 25, no. 1 (June 1938): pp. 48–49.

3. Fominaya, "Collective Identity in Social Movements," p. 394. See also Stets and Burke, "Identity Theory and Social Identity Theory," p. 232.

4. Fominaya, "Collective Identity in Social Movements," pp. 395–398, passim. Gordon H. McCormick and Frank Giordano, "Things Come Together: Symbolic Violence and Guerrilla Mobilisation," *Third World Quarterly* 28, no. 2, The Long War: Insurgency, Counterinsurgency, and Collapsing States (2007): p. 297.

5. Christy Aroopala, "Mobilizing Collective Identity: Frames and Rational Individuals," *Political Behavior* 34, no. 2 (June 2012): pp. 194–197, 199–202.

6. Francesca Polletta and James M. Jasper, "Collective Identity and Social Movements," *Annual Review of Sociology* 27 (2007): pp. 284, 296.

7. Alberto Melucci, "The Process of Collective Identity," in *Social Movements and Culture*, eds. Hank Johnson and Bert Klandermans (Minneapolis: University of Minnesota Press, 1995), pp. 41–63. Marc Howard Ross, "A Cross-Cultural Theory of Political Conflict and Violence," *Political Psychology* 7, no. 3 (September 1986): p. 428.

8. Polletta and Jasper, "Collective Identity and Social Movements," p. 291. Steven Pfaff, "Collective Identity and Informal Groups in Revolutionary Mobilization: East Germany in 1989," *Social Forces* 75, no. 1 (September 1996): pp. 91, 95, 99, 104. Peter du Preez, *The Politics of Identity* (Oxford: Basil Blackwell, 1980), p. 3.

9. John McCardell, *The Idea of a Southern Nation: Southern Nationalists and Southern Nationalism, 1830–1860* (New York: W.W. Norton, 1979), pp. 36–38. Rogers Brubaker, *Ethnicity without Groups* (Cambridge, MA: Harvard University Press, 2004), pp. 47–48. Anne Sarah Rubin, *A Shattered Nation: The Rise and Fall of the Confederacy, 1861–1868* (Chapel Hill: University of North Carolina Press, 2005), pp. 111, 117.

10. Rubin, *Shattered Nation*, pp. 69, 117, 174, 179.

11. Ross, "Cross-Cultural Theory," pp. 434–438.

12. Andrew Mack, "Why Nations Lose Small Wars: The Politics of Asymmetric Conflict," *World Politics* 27, no. 2 (January 1975): pp. 179, 181. du Preez, *The Politics of Identity*, pp. 1, 3.

13. Ted Tunnell, "Creating 'The Propaganda of History': Southern Editors and the Origins of 'Carpetbagger and Scalawag,'" *Journal of Southern History* 72, no. 4 (November 2006): p. 820.

14. Prince, *Stories of the South*, p. 53. *The National Era* (Washington, D.C.), August 24, 1871, has a definition of "carpet-bagger."

15. Thad Dunning, "Fighting and Voting: Violent Conflict and Electoral Politics," *Journal of Conflict Resolution* 55, no. 3 (June 2011): pp. 328–329.

16. Dunning, "Fighting and Voting," pp. 329–330.

17. Gregory P. Downs, *The Declaration of Dependence: The Long Reconstruction of Popular Politics in the South, 1861–1908* (Chapel Hill: University of North Carolina Press, 2011), p. 112.

18. Dunning, "Fighting and Voting," pp. 331, 336.

19. McCormick and Giordano, "Things Come Together," pp. 306–307. George C. Rable, *But There Was No Peace: The Role of Violence in the Politics of Reconstruction* (Athens: University of Georgia Press, 2007), pp. 87–89. Egerton, *Wars of Reconstruction*, pp. 248–253.

20. *The Tri-Weekly Standard* (Raleigh, NC), July 25, 1867, and August 20, 1867.

21. *Anderson Intelligencer* (Anderson Court House, SC), August 28, 1867. The report also appeared in the *Yorkville Courier* (SC), September 12, 1867. See also Thomas B. Alexander, "Political Reconstruction in Tennessee 1865–1870," in *Radicalism, Racism, and Party Realignment: The Border States during Reconstruction*, ed. Richard O. Curry (Baltimore: Johns Hopkins University Press, 1969), pp. 65–66, 70.

22. *The Western Democrat* (Charlotte, NC), November 12, 1867; *The Yorkville Enquirer* (SC), October 24, 1867; *Public Ledger* (Memphis, TN), December 30, 1867; *Nashville Union and Dispatch* (TN), July 21, 1867.

23. *The Pulaski Citizen* (TN), July 19, 1867.

24. *The Louisiana Democrat* (New Orleans, LA), April 29, 1868.

25. Ben H. Severance, *Tennessee's Radical Army: The State Guard and Its Role in Reconstruction, 1867–1869* (Knoxville: University of Tennessee Press, 2005), pp. 6–7, 33, 64–65, 126, 148–154. Severance describes the Tennessee militia as Brownlow's "partisan political army." A newspaper reported that freedmen were given a ballot and told that it represented "forty acres, a mule, freedom, votes, and the equal of the white man." *The Western Democrat* (Charlotte, NC), November 12, 1867.

26. An Act of Congress passed in February 1865 prohibited military interference in elections, unless faced with armed resistance. A commander could take action in keeping order at polling places if the civil authorities were unable to do so.

27. Rable, *But There Was No Peace*, pp. 75–78.

28. Downs, *Declaration of Dependence*, pp. 102, 104.

29. Wayne K. Durrill, "Political Legitimacy and Local Courts: 'Politicks at Such a Rage' in a Southern Community during Reconstruction," *Journal of Southern History* 70, no. 3 (August 2004): pp. 580, 582, 601.

30. Edwards, "Reconstruction and the History of Governance," in *The World the Civil War Made*, eds. Downs and Masur, pp. 24, 28, 32.

31. William C. Harris, "Mississippi Republican Factionalism and Mismanagement," in *Reconstruction and Redemption in the South*, ed. Otto Olsen, p. 81. James D. Fearon and David D. Laitin, "Ethnicity, Insurgency, and Civil War," *American Political Science Review* 97, no. 1 (February 2003): pp. 80–81. Edwards, "Reconstruction and the History of Governance." in *The World the Civil War Made*, eds. Downs and Masur, pp. 29–31, 32, 35.

32. *Weekly North-Carolina Standard* (Raleigh, NC), April 24, 1867.

33. Richard L. Hume, "Carpetbaggers in the Reconstruction South: A Group Portrait of Outside Whites in the 'Black and Tan' Constitutional Conventions," *Journal of American History* 64, no. 2 (September 1977): pp. 316–317, 324. The same problems emerged in West Virginia. The ironclad oath disqualified the majority of whites, achieving the immediate goal of eliminating former Confederates as a political threat. The legislators were far less concerned about the rights of blacks, but with a white minority being able to vote, enfranchisement of blacks was the only way to maintain power and establish a government. The black population in West Virginia was quite small, less than 5 percent; as a result, political battles would continue to rage between Radical and conservative whites over control of the black vote. Even in West Virginia, there were attempts to restrict black votes during elections, but these actions lack the elements of asymmetric warfare. Much of the discontent with black voting was competition with whites over jobs. See Stephen D. Engle, "Mountaineer Reconstruction: Blacks in the Political Reconstruction of West Virginia," *Journal of Negro History* 78, no. 3 (Summer 1993): p. 147.

34. Downs, *Declaration of Dependence*, p. 101.

35. Peter Kolchin, "Carpetbaggers and Reconstruction: A Quantitative Look at Southern Congressional Politics, 1868–1872," *Journal of Southern History* 45, no. 1 (February 1979): pp. 71–76. See also Jeffrey J. Crow, "Thomas Settle, Jr., Reconstruction, and the Memory of the Civil War," *Journal of Southern History* 12, no. 4 (November 1996): pp. 689, 705, 707.

36. Edward Pierce, *Memoir and Letters of Charles Sumner*, vol. 4 (Cambridge, MA: Press of John Wilson and Son, 1877), p. 319.

37. Richard L. Hume, "The Arkansas Constitutional Convention of 1868: A Case Study in the Politics of Reconstruction," *Journal of Southern History* 39, no. 2 (May 1973): pp. 186, 196, 200. Hume notes that the scalawag votes were split between those who fully supported the Radical Republicans and those who did not believe blacks were equal to whites and were concerned that the new constitution would lead to granting blacks social privileges.

38. *Anderson Intelligencer* (Anderson Court House, SC), December 9, 1868.

39. Quoted in the *Memphis Daily Appeal* (TN), September 19, 1867; *The Edgefield Advertiser* (SC), March 25, 1868.

40. *The Louisiana Democrat* (New Orleans, LA), April 29, 1868.

41. *Carolina Watchman* (Salisbury, NC), August 26, 1867.

42. The narrative, as Michael Vlahos defines it, is identity's core meaning, "organized and sacredly packaged into a formal liturgy." This narrative is shaped through the process of collective identity formation. It serves to define reality and provides the commitment and solidarity and sustainment that contribute to the asymmetry of will. See Vlahos, *Fighting Identity*, pp. 78, 138. This categorization, of course, reflects the standard Dunning school approach, but when placed in terms of identity formation, these labels have a powerful meaning in explaining collective identity formation and mobilization and why they have endured. W. Scott Poole, *Never Surrender: Confederate Memory and Conservatism in the South Carolina Upcountry* (Athens: University of Georgia Press, 2004), pp. 53–54, 59, 63. Robert F. Durden, "The Prostrate State Revisited: James K. Pike and South Carolina Reconstruction," *Journal of Negro History* 19, no. 2 (April 1954): pp. 106–107.

43. Tunnell, "Creating 'The Propaganda of History,'" pp. 792, 800. *The Daily Clarion*, Jackson, Mississippi, May 20, 1868.

44. Simpson, "Reconstruction and Reconciliation," pp. 477–478. *The Louisiana Democrat*, New Orleans, Louisiana, April 29, 1868.

CHAPTER 5

1. Stets and Burke, "Identity Theory and Social Identity Theory," pp. 226–228, 230.

2. Elaine Frantz Parsons, "Midnight Rangers: Costume and Performance in the Reconstruction-era Ku Klux Klan," *Journal of American History* 92, no. 3 (December 2005): pp. 815–816.

3. Elaine Frantz Parsons, *Ku-Klux: The Birth of the Klan during Reconstruction*, (Chapel Hill: University of North Carolina Press, 2015), pp. 111–112, 114–115.

4. Parsons, "Midnight Rangers," pp. 832–833. Parsons describes the bizarre appearance of Klan members as "an image of a civilized body capable of harnessing and containing savagery within a reconstructed social world." It was this capability for savagery, Parsons asserts, that contributed to a reassertion of Southern white manhood.

5. Ross, "Cross-Cultural Theory," p. 450. Matthew Hoddie, and Caroline Hartzell, "Signals of Reconciliation: Institution-Building and the Resolution of Civil Wars," *International Studies Review* 7, no. 1 (March 2005): p. 22.

6. Michael W. Fitzgerald, "Property Crime and the Plantation System in Reconstruction Alabama," *Agricultural History Society* 71, no. 2 (Spring 1997): pp. 191, 195.

7. *The Pulaski Citizen* (TN), September 6, 1867.

8. *Nashville Union and Dispatch* (TN), December 16, 1867.

9. *The Athens Post* (TN), February 14, 1868.

10. *The Pulaski Citizen* (TN), May 1, 1868. As an indication of the power of the Klan as an idea, two advertisements appeared in the Pulaski newspaper: one for a land sale authorized by the Grand Cyclops of the Klan, and the other for J.C. Lambeth Company, offering "fancy groceries and confectioneries" for parties, weddings, and picnics, which highlighted a Klan soda fountain. Second quotation is from the *Daily Clarion* (Jackson, MS), March 30, 1868.

11. *The Daily Phoenix* (Columbia, SC), April 28, 1868, and April 15, 1868.

12. *New Orleans Republican* (LA), May 15, 1868; *Wyoming Democrat* (Tunkhannock, PA), July 22, 1868.

13. *Union Flag* (Jonesboro, TN), July 24, 1868, which was quoting from the Unionist *Knoxville Whig*.

14. *The Daily Phoenix* (Columbia, SC), February 23, 1871.

15. J. Michael Martinez, *Carpetbaggers, Cavalry, and the Ku Klux Klan* (New York: Rowman and Littlefield, 2007), p. 71.

16. *Sunbury American* (PA), September 12, 1868.

17. *Lamoille Newsdealer* (Hyde Park, VT), April 5, 1871.

18. During the election for circuit court officers in May of 1869, white Republicans were treated to "all manner of insults and indignity, scorn, contempt, and threats." See Severance, *Tennessee's Radical Army*, p. 224.

19. McCormick and Giordano, "Things Come Together," pp. 311–313, 317–318.

20. Edward John Harcourt, "The Whipping of Richard Moore: Reading Emotion in Reconstruction America," *Journal of Southern History* 36, no. 2 (Winter 2002): p. 261.

21. Kalyvas, "The Ontology of Political Violence," pp. 477–479, passim. See also Richter, "'Shoot or Get Out of the Way!': The Murder of Texas Freedmen's Bureau Agent William G. Kirkman by Cullen Baker—and the Historians," pp. 66–67, and Douglas Kubicek and Carroll Scogin-Brincefield, "An Uncompromising Line between Yankee Rule and Rebel Rowdies: Reconstruction Violence in Lavaca County," in *Still the Arena of Civil War: Violence and Turmoil in Reconstruction Texas, 1865–1874*, ed. Kenneth W. Howell (Denton, TX: University of North Texas Press, 2012), pp. 78–79, 88–89, 380–381. Baker was a classic western desperado. He was a Confederate deserter who killed indiscriminately. Although he and his band of young men were operating at the time of the rise of Klan activity, there appears to be no association. It appears that Baker targeted Kirkman because of his high profile attempts to restrain lawlessness and his status as a Yankee outsider. In 1868, to suppress violence in northeast Texas, U.S. troops conducted a series of unrestrained sweeps to capture gang members (or Klan-affiliated

individuals—no distinction was made). Civilians were subject to military punishments when questioned, and captured suspects were killed outright.

22. Joe Gray Taylor, "Louisiana: An Impossible Task," in *Reconstruction and Redemption in the South*, ed. Otto Olsen (Baton Rouge: Louisiana State University Press, 1980), pp. 205, 226.

23. Robert M. Fogelson and Richard E. Rubenstein, eds., *Use of the Army in Certain Southern States* (New York: Arno Press, 1969), p. 239.

24. Ibid., pp. 239–240.

25. *Anderson Intelligencer* (Anderson Court House, SC), March 4, 1869. House of Representatives, 40th Congress, 3d session, *Message of the President of the United States and Accompanying Documents to the Two Houses of Congress*, Report of the Secretary of War (Washington, D.C.: Government Printing Office, 1868), pp. 467–469. See also the *Daily Phoenix* (Columbia, SC), December 2, 1868, and the *Anderson Intelligencer* (Anderson Court House, SC), December 9, 1868. For the account of the Dill murder, see the *Charleston Daily News* (SC), June 13, 1868. See also Herbert Shapiro, "The Ku Klux Klan during Reconstruction: The South Carolina Episode," *Journal of Negro History* 49, no. 1 (January 1964): pp. 35, 37–39.

26. See Elizabeth Otto Daniell, "The Ashburn Murder Case in Georgia Reconstruction, 1868," *Georgia Historical Quarterly* 59, no. 3 (Fall 1975): pp. 296–312. See also George G. Meade, *Major General Meade's Report on the Ashburn Murder* (Atlanta, GA: U.S. Army Department of the South, 1868). *Weekly North-Carolina Standard* (Raleigh, NC), April 22, 1868. Sefton, *United States Army and Reconstruction*, pp. 170–172. *The New York Herald* (NY), April 8, 1868, criticized Major General Meade for taking action to suppress the Ku-Klux but not equal action to address the Loyal League activities, observing that "the exasperation which will necessarily follow such a one-sided policy will not fail to precipitate a conflict."

27. *Encyclopedia of Arkansas History & Culture*, s.vv. "James Hinds," accessed September 4, 2016, http://www.encyclopediaofarkansas.net.

28. *The New York Herald* (NY), January 18, 1869. *The Athens Post* (GA), March 19, 1869. *The Weekly Standard* (Raleigh, NC), October 6, 1869.

29. *The Weekly Standard* (Raleigh, NC), September 15, 1868.

30. Karin L. Zipf, "'The Whites Shall Rule the Land or Die': Gender, Race, and Class in North Carolina," *Journal of Southern History* 65, no. 3 (August 1999): p. 501. *Nashville Union and Dispatch* (TN), July 24, 1868.

31. Wetherington, *Plain Folk's Fight*, p. 280. Poole, *Never Surrender*, pp. 78, 96–97.

32. Kalyvas, *Logic of Violence in Civil War*, pp. 10, 23, 365.

33. Ibid., 325. Poole, *Never Surrender*, pp. 101–102.

34. Wetherington, *Plain Folk's Fight*, p. 265.

35. Tunnell, "Creating 'The Propaganda of History,'" p. 790.

36. *Nashville Union and American* (TN), April 13, 1870.

37. *The Evansville Journal* (IL), April 25, 1870.

38. *Stark County Democrat* (Canton, OH), April 27, 1870 (first quotation). *New Era* (Washington, D.C.), July 21, 1870 (second quotation).

39. *The Daily Phoenix* (Columbia, SC), April 22, 1870.

40. *New Era* (Washington, D.C.), August 18, 1870.

41. Charles W. Calhoun, *Conceiving a New Republic: The Republican Party and the Southern Question, 1869–1900* (Lawrence: University Press of Kansas, 2006), pp. 14–15, 18–19.

42. *Public Ledger* (Memphis, TN), April 12, 1869. *The Daily Phoenix* (Columbia, SC), April 6, 1870.

43. The fact that Bullock held a commission in the Confederate army and was not disfranchised indicates the machinations behind the process of identifying loyal whites. In the search for acceptable voters, being a Northerner had a great deal of influence. Apparently the voter registration board overlooked Bullock's commission because he was a Northerner who moved South only because his employer moved him there, and he continued his business dealings during the war, even as he was handling affairs for the Confederate Quartermaster Department.

44. *Richmond Dispatch* (VA), September 5, 1868; *Keowee Courier* (Pickens Court House, SC), September 11, 1868; *The Daily Phoenix* (Columbia, SC), September 18, 1868.

45. Digital Library of Georgia, "Civil War Unrest in Camilla, Georgia, 1868," accessed August 17, 2016, http://dlg.galileo.usg.edu/camilla. Sefton, *United States Army and Reconstruction*, pp. 198–199.

46. W. Calvin Smith, "The Reconstruction 'Triumph' of Rufus B. Bullock," *Georgia Historical Quarterly* 52, no. 4 (December 1968): pp. 414–425. Sefton, *United States Army and Reconstruction*, pp. 203–206.

47. *New Era* (Washington, D.C.), April 7, 1870.

48. These were quotations taken from a number of newspapers from around the country that were published in the *New National Era* (Washington, D.C.), October 13, 1870, and in the *Knoxville Weekly Chronicle* (TN), September 21, 1870.

49. Rable, *But There Was No Peace*, pp. 71, 73.

50. Severance, *Tennessee's Radical Army*, p. 180.

51. Ibid., pp. xiii–xiv, 162–163, 185–186.

52. Ibid., pp. xvii, 20, 204–206.

53. Ibid., pp. 224–228.

54. Rable, *But There Was No Peace*, pp. 105–107. Thomas A. DeBlack, *With Fire and Sword: Arkansas, 1861–1974* (Fayetteville: University of Arkansas Press, 2003), p. 191.

55. *Public Ledger* (Memphis, TN), January 8, 1869.

56. For example, see the *Athens Post* (TN), January, 22, 1869; the *Public Ledger* (Memphis, TN), January 8, 1869; the *Cairo Evening Bulletin* (IL), January 14, 1869; *The Southerner* (Tarboro, NC), February 11, 1869; the *Staunton Spectator* (VA), February 9, 1869; and the *Memphis Daily Appeal* (TN), January 29, 1869.

57. *The New Orleans Crescent* (LA), January 13, 1869.

58. DeBlack, *With Fire and Sword*, pp. 181–185, 192–198.

59. A Virginia newspaper immediately compared North Carolina governor Holden to Clayton and Brownlow as Holden sought to raise a militia force. *Staunton Spectator* (VA), November 9, 1869.

60. *The Memphis Daily Appeal* (TN), April 23, 1871.

61. DeBlack, *With Fire and Sword*, pp. 214–218.

62. Sefton, *United States Army and Reconstruction*, pp. 236–239.

63. Jerrell H. Shofner, *Nor Is It Over Yet: Florida in the Era of Reconstruction, 1863–1877* (Gainesville: University Press of Florida, 1974), pp. 195, 205–209, 214–215.

64. Daniel R. Weinfeld, *The Jackson County War: Reconstruction and Resistance in Post-Civil War Florida* (Tuscaloosa: University of Alabama Press, 2012), pp. 31, 34–35.

65. Ibid.

66. Ibid., pp. 66–67, 92–93.

67. *Richmond Dispatch* (VA), December 23, 1869.

68. Shofner, *Nor Is It Over Yet*, pp. 228–235. Allen W. Trelease, *White Terror: The Ku Klux Klan Conspiracy and Southern Reconstruction* (New York: Harper and Row, 1971), pp. 119–120, 311–312. Weinfeld, *Jackson County War*, pp. 112–117.

69. Sefton, *Army and Reconstruction*, pp. 234–235.

70. Weinfeld, *Jackson County War*, pp. 102–105, 109–111, 125.

71. Shofner, *Nor Is It Over Yet*, pp. 200–205, 208–210, 218–221.

72. Canter Brown, Jr., *Ossian Bingley Hart: Florida's Loyalist Reconstruction Governor* (Baton Rouge: Louisiana State University Press, 1997), pp. 282–284.

73. Shofner, *Nor Is It Over Yet*, pp. 263–266.

CHAPTER 6

1. *The New York Herald* (NY), March 9, 1871. *New National Era* (Washington, D.C.), October 5, 1871. Harris, "Mississippi: Republican Factionalism and Mismanagement," pp. 80–81.

2. *The Highland Weekly News* (Hillsboro, OH), September 28, 1871. *The Jeffersonian* (Stroudsburg, PA), August 10, 1871. Select Committee of the Senate, 42d Congress, 1st session, *A Report on the Alleged Outrages in the Southern States*, March 10, 1871 (Washington, D.C.: U.S. Government Printing Office, 1871).

3. *Anderson Intelligencer* (Anderson Court House, SC), May 25, 1871.

4. Fogelson and Rubenstein, eds., *Use of the Army in Certain Southern States*, p. 9.

5. Ibid. See also Francis B. Simkins, "The Ku Klux Klan in South Carolina, 1868–1871," *Journal of Negro History* 12, no. 4 (October 1927): p. 641.

6. *The Daily State Register* (Carson City, NV), April 25, 1871; *The States and Union* (Ashland, OH), May 17, 1871; *The Evening Telegraph* (Philadelphia, PA), June 23, 1871.

7. The testimonies collected fills 13 volumes. Allen Trelease and others have used the testimonies of high-profile individuals, such as Nathan Bedford Forrest and John B. Gordon, who were less than forthcoming and quite vague before the committee, to represent Southerners in general as guilty prevaricators. In actuality, a review of the index and contents of each volume of testimony for every state reveals a remarkably consistent and strikingly honest response of white Southerners to the source of resistance and violence in their states that represents the mobilization of a collective identity. See Trelease, *White Terror*, pp. 393–397.

8. Thomas A. Scott, ed., *Cornerstones of Georgia History: Documents the Formed the State* (Athens: University of Georgia Press, 1995), p. 116.

9. Parsons, *Birth of the Klan during Reconstruction*, pp. 127–130, 133.

10. *Anderson Intelligencer* (Anderson Court House, SC), February 29, 1872. *The Daily Phoenix* (Columbia, SC), April 26, 1872.

11. *The New York Herald* (NY), September 15, 1871.

12. Foner, *Reconstruction*, p. 457. Sefton, *Army and Reconstruction*, p. 222.

13. *The South-Western* (Shreveport, LA), February 9, 1870.

14. Simkins, "The Ku Klux Klan in South Carolina," pp. 608, 611, 633–635. Martinez, *Carpetbaggers, Cavalry, and the Ku Klux Klan*, p. 111.

15. Simkins, "The Ku Klux Klan in South Carolina," pp. 608, 611, 633–635. *The Fairfield Herald* (Winnsboro, SC), March 1, 1871. Shapiro, "The Ku Klux Klan during Reconstruction," pp. 39–43.

16. *The True Northerner* (Paw Paw, MI), October 18, 1872.

17. Martinez, *Carpetbaggers, Cavalry, and the Ku Klux Klan*, pp. 135, 138–139, 140, 145, 151. Stanley F. Horn, *Invisible Empire: The Story of the Ku Klux Klan, 1866–1871*, 2d ed. (Montclair, NJ: Patterson, Smith Publishing Corporation, 1939), pp. 238–239.

18. Lou Falkner Williams, *The Great South Carolina Ku Klux Klan Trials, 1871–1872* (Athens: University of Georgia Press, 1996), pp. 82–84.

19. Williams, *South Carolina Ku Klux Klan Trials*, pp. 120–121. Horn, *Invisible Empire*, pp. 240–241.

20. James A. Rawley, "The General Amnesty Act of 1872: A Note," *Mississippi Valley Historical Review* 47, no. 3 (December, 1960): pp. 480, 482. See also Paul H. Buck, *The Road to Reunion 1865–1900* (New York: Vintage Books, 1937), pp. 128–131.

21. *Wilmington Journal* (NC), September 13, 1872.

22. *The New York Herald* (NY), December 14, 1872.

23. Williams, *South Carolina Ku Klux Klan Trials*, pp. 124–125. Williams makes a grim effort to validate the results of the trials, pointing out that the guilty were punished, and asserting that the white South was convinced by these trials "that the federal government would no longer tolerate the counterrevolution the Ku Klux Klan represented" (p. 122).

CHAPTER 7

1. Downs, *Declaration of Dependence*, p. 106. Zipf, "'Whites Shall Rule the Land or Die,'" pp. 504–506.

2. Olsen, "North Carolina: An Incongruous Presence," pp. 152, 172.

3. Scott Reynolds Nelson, *Iron Confederacies: Southern Railways, Klan Violence and Reconstruction* (Chapel Hill: University of North Carolina Press, 1998), pp. 100, 107, 110, 114. Nelson argues that the railroad in Alamance County served as a center of political violence. See also Trelease, *White Terror*, pp. 200–201.

4. *The Weekly Standard* (Raleigh, NC), April 27, 1870.

5. Olsen, "North Carolina: An Incongruous Presence," pp. 179–180. Crow, "Thomas Settle, Jr., Reconstruction and the Memory of the Civil War," pp. 708–709.

The Weekly Standard (Raleigh, NC), July 27, 1871. The governor issued a proclamation offering a $500 reward for the arrest of the murderers. See the *Weekly Standard* (Raleigh, NC), June 1, 1870, for Stephens's background and account of his murder. See the *Wilmington Journal* (NC), August 26, 1870; *The Southerner* (Tarboro, NC), August 25, 1870; and the *Weekly Standard* (NC), August 31, 1870, for affidavits and testimony of accusers and accused. Trelease, *White Terror*, pp. 214–215. A detailed account of Outlaw's life and death is found in Nelson, *Iron Confederacies*. See also Horn, *Invisible Empire*, pp. 197–198.

6. *The Weekly Standard* (Raleigh, NC), September 7, 1870.

7. *The Weekly Standard* (Raleigh, NC), August 3, 1870.

8. Crow, "Thomas Settle, Jr., Reconstruction and the Memory of the Civil War," pp. 708–709.

9. *New Era* (Washington, D.C.), August 18, 1870. See also the *Weekly Standard* (Raleigh, NC), August 3, 1870, for a survey of Ku-Klux activities and members, described as "gentlemen of high family and standing." See also the *Weekly Standard* (Raleigh, NC), July 27, 1870, and June 6, 1870.

10. Horn, *Invisible Empire*, pp. 210–212. Trelease, *White Terror*, pp. 224–225.

11. *The Charlotte Democrat* (NC), March 21, 1871.

12. Crow, "Thomas Settle, Jr., Reconstruction, and the Memory of the Civil War," p. 709.

13. Patrick W. Riddleberger, "The Radicals' Abandonment of the Negro during Reconstruction," *Journal of Negro History* 45, no. 2 (April 1960): pp. 91–93.

14. Michael F. Holt, *By One Vote: The Disputed Presidential Election of 1876* (Lawrence: University Press of Kansas, 2008), pp. 8–11. Michael W. Fitzgerald, *Splendid Failure: Postwar Reconstruction in the American South*, The American Ways Series (Chicago: Ivan R. Dee, 2007), pp. 175–177.

15. Taylor, "Louisiana: An Impossible Task," pp. 205, 226.

16. Fogelson and Rubenstein, eds., *Use of the Army in Certain Southern States*, p. 344–345.

17. Ibid., p. 354.

18. Ibid., p. 163.

19. Ibid., pp. 219–220.

20. Ibid., p. 230.

21. Ibid., p. 223.

22. Ibid., pp. 255, 260–261.

23. Joseph G. Dawson, III, *Army Generals and Reconstruction, 1862–1877* (Baton Rouge: Louisiana State University Press, 1982), pp. 136–142.

24. James K. Hogue, *Uncivil War: Five New Orleans Street Battles and the Rise and Fall of Radical Reconstruction* (Baton Rouge: Louisiana State University Press, 2006), pp. 103–106.

25. Charles Lane, *The Day Freedom Died: The Colfax Massacre, the Supreme Court, and the Betrayal of Reconstruction* (New York: Henry Holt, 2008), pp. 85–109. See also the *New Orleans Bulletin* (LA), February 13, 1875.

26. Fogelson and Rubenstein, eds., *Use of the Army in Certain Southern States*, p. 386. *The Shreveport Times* (LA), July 4, 1874. Italics in the original.

27. *The New Orleans Bulletin* (LA), December 10, 1874.

28. *The Opelousas Courier* (St. Landry Parish, LA), August 15, 1875.

29. Fogelson and Rubenstein, eds., *Use of the Army in Certain Southern States*, p. 375. *The Shreveport Times* (LA), July 18, 1874. Italics in the original.

30. Ibid., p. 262.

31. Ibid., pp. 356–357. See also Hogue, *Uncivil War*, p. 126.

32. Ibid., p. 194.

33. Ibid., pp. 356–357. See also Hogue, *Uncivil War*, pp. 128–132.

34. Hogue, *Uncivil War*, pp. 136–137.

35. Sefton, *Army and Reconstruction*, pp. 239–240.

36. Joe Gray Taylor, *Louisiana Reconstructed, 1863–1877* (Baton Rouge: Louisiana State University Press, 1974), pp. 294–296; Hogue, *Uncivil War*, pp. 138–142; Budiansky, *Bloody Shirt*, p. 176; Dawson, *Army Generals and Reconstruction: Louisiana*, pp. 167–180.

37. *New Orleans Republican* (LA), January 23, 1875. See also the *Nashville Union and American* (TN), January 1, 1875; the *New Orleans Bulletin* (LA), February 3, 1875. *The Daily Dispatch* (Richmond, VA), January 2, 1875.

38. Fogelson and Rubenstein, eds., *Use of the Army in Certain Southern States*, pp. 382–383. *The Boston Daily Globe*, Boston, Massachusetts, September 23, 1874. See the *New Orleans Bulletin* (LA), October 11, 1874, for a scathing reply to Jewett, who was reported missing after being charged with embezzlement. *New York Tribune* (NY), October 31, 1874.

39. William C. Harris, *The Day of the Carpetbagger: Republican Reconstruction in Mississippi* (Baton Rouge: Louisiana State University Press, 1979), pp. 609, 621–622.

40. *The Opelousas Courier* (St. Landry Parish, LA), August 15, 1875.

41. *The New Orleans Bulletin* (LA), December 10, 1874. Harris, *Republican Reconstruction in Mississippi*, p. 646.

42. *Daily Alta California* (San Francisco, CA), December 17, 1874.

43. Ibid., p. 21. Reprinted from the *Commercial* (Cincinnati, OH). Ibid., December 17, 1874.

44. House of Representatives, State of Mississippi, *The Testimony in the Impeachment of Adelbert Ames as Governor of Mississippi* (Jackson, Mississippi: Power and Barksdale, State Printers, 1877), pp. 5, 250–260. Vicksburg Meridian *Mercury* as reported in the *Jackson Times* (MS), (June 8, 1875.

45. *The Weekly Clarion* (Jackson, MS), January 21, 1875.

46. *The Indiana State Sentinel* (Indianapolis, IN), June 17, 1875. *The Bossier Banner* (Bellevue, LA), March 13, 1875. See also the *Democratic Press* (Ravenna, OH), April 1, 1875; the *Louisiana Democrat* (Alexandria, LA), March 17, 1875; and the *Helena Weekly Herald* (MT), January 19, 1875. See also Olsen, ed. *Reconstruction and Redemption in the South*, pp. 97, 99–100. Budiansky, *Bloody Shirt*, p. 193.

47. Nicholas Lehman, *Redemption: The Last Battle of the Civil War* (New York: Farrar, Straus and Giroux, 2006), pp. 100–103.

48. Harris, *Republican Reconstruction in Mississippi*, p. 662. *The Weekly Echo* (Lake Charles Parish, Calcasieu, LA), September 16, 1875. *The Greenville Times* (MS), September 11, 1875. *The Weekly Clarion* (Jackson, MS), September 8, 1875.

49. Assembled from accounts printed in various newspapers, especially the *Weekly Clarion* (Jackson, MS), September 8, 1875; the *National Republican* (Washington, D.C.), September 9, 1875; and the *Memphis Daily Appeal* (TN), September 8, 1875. *The Bolivar Bulletin* (TN), September 10, 1875.

50. Olsen, ed., *Reconstruction and Redemption in the South*, pp. 102–103.

51. *The Daily Clarion* (Jackson, MS), October 11, 1875. *The Weekly Clarion* (Jackson, MS), October 13, 1875. Harris, *Republican Reconstruction in Mississippi*, p. 662.

52. Fitzgerald, *Splendid Failure*, pp. 191–192. Olson, ed. *Reconstruction and Redemption in the South*, pp. 100, 104–106. Harris, *Republican Reconstruction in Mississippi*, pp. 666–667, 672–673.

53. House of Representatives, State of Mississippi, *The Testimony in the Impeachment of Adelbert Ames as Governor of Mississippi* (Jackson, Mississippi: Power and Barksdale, State Printers, 1877), pp. 5, 250–260. Olsen, ed., *Reconstruction and Redemption in the South*, pp. 107–108.

54. *The New Orleans Bulletin* (LA), January 14, 1875.

55. Dawson, *Army Generals and Reconstruction*, pp. 52–60. *Forest Republican* (Tionesta, PA), January 13, 1875.

56. *The Daily Dispatch* (Richmond, VA), January 2, 1875.

57. Hogue, *Uncivil War*, pp. 148–150. Sefton, *U.S. Army and Reconstruction*, pp. 240–242. Rable, *But There Was No Peace*, pp. 141–142. See the *Indiana State Sentinel* (Indianapolis, IN), January 5, 1875, for a detailed summary of the background to the decision for dispatching Sheridan. Taylor, *Louisiana Reconstructed*, pp. 304–308.

58. See the *New York Herald* (NY), January 6, 1875, which printed copies of the dispatches. Sheridan may not have been far from wrong. A Donaldsville, Louisiana, newspaper reported that "recent reports from the neighborhood of Shreveport show that the whole section of the country has fallen into a kind of anarchy." *The Donaldsville Chief* (LA), January 16, 1875. See also the *Helena Weekly Herald* (MT), January 7, 1875. Maj. Lewis Merrill told a Congressional committee in early 1875 that "the White League is the only power in the State." Ted Tunnell, *Crucible of Reconstruction: War, Radicalism and Race in Louisiana 1862–1877* (Baton Rouge: Louisiana State University Press, 1984), p. 204.

59. *The New Orleans Bulletin* (LA), January 14, 1875. Sheridan reported that more than 4,200 people had been killed or injured in Louisiana as a result of political violence. See the *New Orleans Republican* (LA), February 9, 1875.

60. *The Daily Argus* (Rock Island, IL), January 12, 1875.

61. *The Daily Dispatch* (Richmond, VA), January 13, 1875. Italics in original. The *New Orleans Republican* (LA), in its January 17, 1875, issue, sought to counter the outrage by publishing three columns of excerpts from newspapers that had expressed support for Sheridan's actions.

62. *The New York Times* (NY), January 8, 1875, as published in the *Los Angeles Daily Herald* (CA), January 10, 1875.

63. *The New York Herald* (NY), January 9, 1875. *The Centre Reporter* (PA), January 21, 1875.

64. *New York Tribune* (NY), January 14, 1875.

65. *Alexandria Gazette* (VA), January 7, 1875.

66. *The Aegis and Intelligencer* (Bel Air, MD), January 15, 1875.

67. *The New Orleans Bulletin* (LA), January 5, 1875. Italics in original. See also Ibid., January 14, 1875. The White League was an "open military association for defensive purposes alone." The people were, the newspaper claimed, "at the mercy of characterless scalawags and ignorant and corrupt negroes."

68. *Alexandria Gazette* (VA), January 7, 1875. Gordon's remarks were made on January 6.

69. Holt, *By One Vote*, p. 21. Taylor, *Louisiana Reconstructed*, pp. 308–310. See also the *New Orleans Bulletin* (LA), February 26, 1875.

70. Foner, *Reconstruction*, pp. 555–556. Andrew L. Slap, *Doom of Reconstruction* (New York: Fordham University Press, 2006), pp. 224–226.

71. First quotation from Fogelson and Rubenstein, eds., *Use of the Army in Certain Southern States*, pp. 133–134. See also a letter from Governor Davis to Attorney General Taft, September 9, 1876, pp. 142–143, and details provided by a witness, pp.144–146. *The New Orleans Bulletin* (LA), February 2, 1875. Last quotation is from the *New Orleans Bulletin* (LA), January 14, 1875.

72. *The Shreveport Times* (LA), September 14, 1874, reprinted from the *New York Herald*. Italics in original.

73. Wetherington, *Plain Folk's Fight*, p. 286, passim. Tunnell, *Crucible of Reconstruction*, p. 193. Jesse T. Carpenter, *The South as a Conscious Minority, 1789–1861: A Study in Political Thought* (Columbia: University of South Carolina Press, 1930), pp. 5, 29, 33, 79–82, 239.

CHAPTER 8

1. Jerry L. West, *The Bloody South Carolina Election of 1876: Wade Hampton III, the Red Shirt Campaign for Governor and the End of Reconstruction* (Jefferson, NC: McFarland, 2011), p. 104.

2. *The Indiana State Sentinel* (Indianapolis, IN), January 17, 1876.

3. *Los Angeles Daily Herald* (CA), May 28, 1876.

4. *The Democratic Press* (Ravenna, OH), August 21, 1876. Similar opinions are found in the *Weekly Register* (Point Pleasant, WV), September 21, 1876; the *Daily Argus* (Rock Island, IL), September 29, 1876; and the *Hartford Herald* (KY), November 1, 1876. *The News and Herald* (Winnsboro, NC), December 21, 1876.

5. *Public Ledger* (Memphis, TN), November 29, 1876. The newspaper defined "Bull-dozer" as "a slang term from Louisianan in the sense of intimidation or violence exercised to influence or control voters and originated among the negroes."

6. *The New York Times* (NY), December 16, 1876. *Ashtabula Telegraph* (OH), August 18, 1876. The *Democratic Advocate* (Westminster, MD), November 25, 1876, reported that "the term 'bulldozers', which is so variously printed in the New Orleans dispatches, is the name applied to an organization of armed white men, whose ostensible business it is to keep the negroes from stealing the cotton crop."

7. *The Somerset Herald* (PA), May 30, 1877. *New Orleans Republican* (LA), October 20, 1876.

8. West, *The Bloody South Carolina Election of 1876*, pp. 67–71. The *Cecil Whig* (Elkton, MD), December 30, 1876, in its account of the Hamburg violence, explicitly associated Gen. M. C. Butler with the bulldozers.

9. Stampp, *Era of Reconstruction*, p. 189.

10. Foner, *Reconstruction*, p. 487.

11. Stampp, *Era of Reconstruction*, pp. 190–191. Slap, *Doom of Reconstruction*, p. 89, 116. *The New York Herald* (NY), April 10, 1876.

12. Heather Cox Richardson, *The Death of Reconstruction: Race, Labor, and Politics in the Post-Civil War North, 1865–1901* (Cambridge: Harvard University Press, 2001), pp. 82, 91, 119, 224.

13. Ibid. *The Columbia Freeman* (Ebensburg, PA), October 20, 1876. Foner, *Reconstruction*, p. 499.

14. *The Bosssier Banner* (Bellevue, LA), October 5, 1876.

15. *The New York Herald* (NY), September 6, 1876.

16. Fogelson and Rubenstein, eds., *Use of the Army in Certain Southern States*, pp. 135, 137.

17. Ibid., pp. 120–121, 147–148, 151.

18. *The New York Herald* (NY), September 6, 1876.

19. Sefton, *Army and Reconstruction*, p. 247.

20. *The Daily Gazette* (Wilmington, DE), August 18, 1876. *Clearfield Republican* (PA), September 27, 1876.

21. West, *The Bloody South Carolina Election of 1876*, p. 87. Rod Andrew, Jr., *Wade Hampton: Confederate Warrior to Southern Redeemer* (Chapel Hill: University of North Carolina Press, 2008), pp. 379, 389–391. Andrew notes that "the Hampton campaign captured the deepest hopes of white South Carolinians and revealed what they had come to believe about themselves" (p. 383).

22. *The Yorkville Enquirer* (SC), October 12, 1876. *The Cecil Whig* (Elkton MD), December 30, 1876.

23. *Anderson Intelligencer* (Anderson Court House, SC), October 20, 1876.

24. *The New York Herald*, October 19, 1876. Sefton, *Army in Reconstruction*, p. 247.

25. *Gallipolis Journal* (OH), October 20, 1876. *Stark County Democrat* (Canton, OH), October 10, 1876.

26. *Anderson Intelligencer* (Anderson Court House, SC), October 26, 1876. Although the committee was adamant that the rifle clubs were not subject to political control, the announcement that the clubs had all been disbanded and officers were "no longer exercising powers," indicated some existing level of command or coordination.

27. The editorial was reprinted in the *Daily Argus* (Rock Island, IL), October 31, 1876.

28. *Clearfield Republican* (PA), September 27, 1876. *Anderson Intelligencer* (Anderson Court House, SC), October 26, 1876.

29. *The News and Herald* (Winnsboro, NC), December 21, 1876.

30. Calhoun, *Conceiving a New Republic*, pp. 98, 100.

31. Holt, *By One Vote*, pp. 124, 130. Quoted in the *Vermont Phoenix* (Battleboro, VT), November 3, 1876.

32. Holt, *By One Vote*, pp. 97–99, 136.

33. Calhoun, *Conceiving a New Republic*, p. 105.

34. *The Weekly Register* (Point Pleasant, WV), September 21, 1876 (first quote). The meeting of the veterans was reported in the *Daily Argus* (Rock Island, IL), September 29, 1876.

35. *The Hartford Herald* (KY), November 1, 1876.

36. *Fayette County Herald* (Washington Court House, OH), July 6, 1876. *New Orleans Republican* (LA), August 17, 1876.

37. Fogelson and Rubenstein, eds., *Use of the Army in Certain Southern States*, pp. 152–153. Roy Morris, Jr., *Fraud of the Century: Rutherford B. Hayes, Samuel Tilden, and the Stolen Election of 1876* (New York: Simon and Schuster, 2003), pp. 131, 149. *New Orleans Republican* (LA), December 28, 1876.

38. *Jackson Standard* (Jackson Court House, OH), November 30, 1876.

39. Ibid.

40. *The Democratic Advocate* (Westminster, MD), November 25, 1876.

41. *The Daily Dispatch* (Richmond, VA), November 20, 1876. *Weekly Democratic Statesman* (Austin, TX), November 23, 1876.

42. Holt, *By One Vote*, pp. 166–167.

43. *The New York Herald* (NY), November 29, 1876. *The True Northerner* (Paw Paw, MI), April 6, 1877. *The Leavenworth Weekly Times* (KS), December 14, 1876.

44. Shofner, *Nor Is It Over Yet*, p. 215, 300–301, 308–313. Jerrel H. Shofner, "Florida: A Failure of Moderate Republicanism," in *Reconstruction and Redemption in the South*, ed. Otto Olsen (Baton Rouge: Louisiana State University Press, 1980), pp. 34–37.

45. William H. Rehnquist, *Centennial Crisis: The Disputed Election of 1876* (New York: Alfred A. Knopf, 2004), pp. 168–178, passim. Holt, *By One Vote*, p. 237. *The Memphis Daily Appeal* (TN), February 24, 1877.

46. Andrew, *Wade Hampton*, pp. 418–419. Morris, *Fraud of the Century*, p. 249. *Orangeburg News and Times* (SC), April 14, 1877. *Anderson Intelligencer* (Anderson Court House, SC), April 5, 1877. Ronald F. King, "Counting the Votes: South Carolina's Stolen Election of 1876," *Journal of Interdisciplinary History* 32, no. 2 (Autumn 2001): pp. 171–172, 179, 187, 190–191.

47. *The News and Herald* (Winnsboro, SC), April 14, 1877, quoted from the *New York Herald*.

48. Dawson, *Army Generals and Reconstruction*, pp. 238–240.

49. Taylor, *Louisiana Reconstructed*, pp. 496–499.

50. Dawson, *Army Generals and Reconstruction*, pp. 246–250. *New Orleans Republican* (LA), March 7, 1877. Armistead L. Robinson, "The Politics of Reconstruction," *Wilson Quarterly* 2, no. 2 (Spring 1978): pp. 106–123.

51. The *Chicago Times* correspondent's report was reprinted in the *Cairo Bulletin* (IL), April 28, 1877.

52. *The New Orleans Daily Democrat* (LA), April 21, 1877. *The Morristown Gazette* (TN), April 25, 1877. Sefton, *Army and Reconstruction*, pp. 248–251. Dawson, *Army Generals and Reconstruction*, pp. 234–242, passim. Hogue, *Uncivil War*, pp. 168–174, passim. Ari Hoogenboom, *The Presidency of Rutherford B. Hayes* (Lawrence: University Press of Kansas, 1988), p. 67.

53. *The New York Times* (NY), December 16, 1877.

54. *The Somerset Herald* (PA), May 30, 1877.

55. *The New Orleans Daily Democrat* (LA), April 25, 1877.

56. *The Somerset Herald* (PA), May 9, 1877. *The Indiana State Sentinel* (Indianapolis, IN), April 25, 1877.

57. *The Cairo Bulletin* (IL), April 28, 1877.

58. *The Newberry Herald* (SC), December 19, 1877 (first quote). See also the *Western Sentinel* (Winston-Salem, NC), October 11, 1877. *Jackson Pilot* (MS), October 4, 1875 (second quote). See also Olsen, ed., *Reconstruction and Redemption in the South*, p. 107.

59. Michael Perman, *Struggle for Mastery: Disfranchisement in the South 1888–1908* (Chapel Hill: University of North Carolina Press, 2001), pp. 10–11. *Jasper Weekly Courier* (IN), August 17, 1877.

60. Matthew Hoddie and Caroline Hartzell, "Signals of Reconciliation: Institution-Building and the Resolution of Civil Wars," *International Studies Review* 7, no. 1 (March 2005): p. 30.

61. *The New York Herald* (NY), May 17, 1876.

62. Buck, *Road to Reunion*, pp. 104–105.

63. Fogelson and Rubenstein, eds., *Use of the Army in Certain Southern States*, p. 138.

64. Karl R. DeRouen Jr. and David Subek, "The Dynamics of Civil War Duration and Outcome," *Journal of Peace Research* 41, no. 3 (May 2004): p. 306.

65. Alastair Smith and Allan C. Stam, "Bargaining and the Nature of War," *Journal of Conflict Resolution* 48, no. 6 (December 2004): pp. 783–784.

CHAPTER 9

1. Ivan Arreguin-Toft, "How the Weak Win Wars: A Theory of Asymmetric Conflict," *International Security*, 26 no. 1 (Summer 2001), pp. 93–95. See also Mack, "Why Big Powers Lose Small Wars," pp. 126–151.

2. Yaacov Bar-Siman-Tov, ed., *From Conflict Resolution to Reconciliation* (New York: Oxford University Press, 2004), pp. 4–5.

3. Daniel Bar-Tal and Gemma H. Bennick, "The Nature of Reconciliation as an Outcome and as a Process," in *From Conflict Resolution to Reconciliation*, ed. Yaacov Bar-Siman-Tov (New York: Oxford University Press, 2004), pp. 13, 15, 17–18, 20–21, 25, 27–28, 35. Marc Howard Ross, "Ritual and the Politics of Reconciliation," in *From Conflict Resolution to Reconciliation*, ed. Bar-Siman-Tov (New York: Oxford University Press, 2004), pp. 36–37.

4. Herbert C. Kelman, "Reconciliation as Identity Change: A Social-Psychological Perspective," in *From Conflict Resolution to Reconciliation*, ed. Yaacov Bar-Siman-Tov (New York: Oxford University Press, 2004), pp. 119–120. Spinner-Halev and Theiss-Morse, "National Identity and Self-Esteem," pp. 523, 525.

5. Calhoun, *Conceiving a New Republic*, pp. 147–156, passim. Kenneth E. Davison, *The Presidency of Rutherford B. Hayes* (Westport, CT: Greenwood Press, 1972), pp. 45, 141–142. Hoogenboom, *Presidency of Rutherford B. Hayes*, p. 205.

6. Buck, *Road to Reunion*, pp. 140–141.

7. John Pettegrew, "The Soldier's Faith: Turn-of-the-Century Memory of the Civil War and the Emergence of Modern American Nationalism," *Journal of Contemporary History* 31 (1996): p. 50.

8. Ibid., pp. 51, 54, 58.

9. Virginius Dabney, *Virginia: the New Dominion* (New York: Doubleday, 1971), p. 237.

10. See William Harris Bragg, "Reconstruction," in *The Civil War in Georgia: A New Encyclopedia Companion*, ed. John C. Inscoe (Athens: University of Georgia Press, 2011), pp. 185–194. Edward L. Ayers, *The Promise of the New South: Life after Reconstruction* (New York: Oxford University Press, 1992), pp. 324–325.

11. Alwyn Barr, "The Black Militia of the New South: Texas as a Case Study," *Journal of Negro History* 63, no. 3 (July 1978): p. 209.

12. Pettegrew, "The Soldier's Faith," pp. 57, 63, 69.

13. David W. Blight, *Beyond the Battlefield: Race, Memory, and the American Civil War* (Amherst: University of Massachusetts Press, 2002), p. 97.

14. Richard K. Herrmann, Pierangelo Isernia, and Paolo Segatti, "Attachment to the Nation and International Relations: Dimensions of Identity and their Relationship to War and Peace," *Political Psychology* 30, no. 5 (October 2009): pp. 723–725. Wiseman Chirwa, "Collective Memory and the Process of Reconciliation and Reconstruction, *Development in Practice* 7, no. 4 (November 1997): p. 482. Buck, *Road to Reunion*, pp. 283–284.

15. The *New York Herald* article was reprinted in the *Cairo Bulletin* (IL), October 6, 1877.

16. C. Vann Woodward, *The Origins of the New South* (Baton Rouge: Louisiana State University Press, 1951), p. 148. *The Daily Phoenix* (Columbia, SC), July 25, 1869. This prescient observation was reprinted from the *New York World*. See also Paul M. Gaston, *The New South Creed: A Study in Southern Mythmaking* (New York: Alfred A. Knopf, 1970), p. 32.

17. Woodward, *Origins of the New South*, pp. 114, 116, 120.

18. Daniel Bar-Tal, "From Intractable Conflict through Conflict Resolution to Reconciliation: Psychological Analysis," *Political Psychology* 21, no. 2 (June 2000): p. 356.

19. Michael O'Brien, *The Idea of the American South, 1920–1941* (Baltimore: Johns Hopkins University Press, 1979), p. 6.

20. See Woodward, *Origins of the New South*, p. 158, and Gaston, *New South Creed*, p. 125.

21. *The Pulaski Citizen* (TN), September 16, 1886. See also Woodward, *Origins of the New South*, p. 151.

22. See the *St. Paul Daily Globe* (MN), December 28, 1886, and the *Abbeville Press and Banner* (SC), December 29, 1886. See also Woodward, *Origins of the New South*, pp. 146–147; Gaston, *New South Creed*, pp. 194–195; and Ayers, *Promise of the New South*, pp. 20–21.

23. Sharon D. Kennedy-Nolle, *Writing Reconstruction: Race, Gender and Citizenship in the Postwar South* (Chapel Hill: University of North Carolina Press, 2015), pp. 282, 284, 292–293.

CONCLUSION

1. T. V. Paul, *Asymmetric Conflicts: War Initiation by Weaker Powers* (New York: Cambridge University Press, 1994), pp. 16, 29.

2. Sullivan, "War Aims and War Outcomes," pp. 499, 501–502, 505.

3. Kathleen Gallagher Cunningham, "Actor Fragmentation and Civil War Bargaining: How Internal Divisions Generate Civil Conflict," *American Journal of Political Science* 57, no. 3 (July 2013): pp. 659–661.

4. Calhoun, *Conceiving a New Republic*, p. 3.

5. *The Weekly Standard* (Raleigh, NC), July 20, 1870.

6. Calhoun, *Conceiving a New Republic*, pp. 6–7. Peter Siani-Davies and Stefanos Katsikas, "National Reconciliation after Civil War: The Case of Greece," *Journal of Peace Research* 46, no. 4 (July 2009): pp. 566, 573.

7. Findley and Edwards, "Weak-Actor Social Structure in Asymmetric Wars," pp. 565–566.

8. Daniel Bar-Tal, "From Intractable Conflict through Conflict Resolution to Reconciliation: Psychological Analysis," *Political Psychology* 21, no. 2 (June 2000): pp. 352–355.

9. This is my tribute to Edward L. Ayers and his deeply evocative portrayal of the New South. See Ayers, *Promise of the New South*, pp. ix–x.

Bibliography

BOOKS AND PERIODICALS

Adams, Gordon, and Shoon Murray, eds. *Mission Creep: The Militarization of U.S. Foreign Policy.* Washington, D.C.: Georgetown University Press, 2014.

Ancker, Clinton J., and Michael D. Burke. "Doctrine for Asymmetric Warfare." *Military Review* 83, no. 4 (July-August 2003): 18–25.

Andrew, Rod, Jr. *Wade Hampton: Confederate Warrior to Southern Redeemer.* Chapel Hill: University of North Carolina Press, 2008.

Arnold, Isaac N. *The Life of Abraham Lincoln.* Chicago: A. C. McClung, 1909.

Aroopala, Christy. "Mobilizing Collective Identity: Frames and Rational Individuals." *Political Behavior* 34, no. 2 (June 2012): 193–224.

Arreguin-Toft, Ivan. "How the Weak Win Wars: A Theory of Asymmetric Conflict." *International Security* 26, no. 1 (Summer 2001): 93–128.

Arreguin-Toft, Ivan. *How the Weak Win Wars: A Theory of Asymmetric Conflict.* New York: Cambridge University Press, 2005.

Auerbach, Yehudith. "The Reconciliation Pyramid: A Narrative-Based Framework for Analyzing Identity Conflicts." *Political Psychology* 30, no. 2 (April 2009): 291–318.

Ayers, Edward L. *The Promise of the New South: Life After Reconstruction.* New York: Oxford University Press, 1992.

Ball-Rokeach, S. J. "Normative and Deviant Violence from a Conflict Perspective." *Social Problems* 28, no. 1 (October 1980): 45–62.

Barr, Alwyn. "The Black Militia of the New South: Texas as a Case Study." *Journal of Negro History* 63, no. 3 (July 1978): 209–19.

Bar-Siman-Tov, Yaacov, ed. *From Conflict Resolution to Reconciliation.* New York: Oxford University Press, 2004.

Bar-Tal, Daniel. "From Intractable Conflict through Conflict Resolution to Reconciliation: Psychological Analysis." *Political Psychology* 21, no. 2 (June 2000): 351–65.

Benford, Robert D., and David A. Snow. "Framing Processes and Social Movements: An Overview and Assessment." *Annual Review of Sociology* 26 (2000): 611–39.

Berger, Mark, and Douglas Borer. "The Long War: Insurgency, Counterinsurgency, and Collapsing States." *Third World Quarterly* 28, no. 2 (March 2007): 197–215.

Bergesen, Albert. "Nation-Building and Constitutional Amendments: The Role of the Thirteenth, Fourteenth, and Fifteenth Amendments in the Legal Reconstitution of the American Polity Following the Civil War." *Pacific Sociological Review* 24, no. 1 (January, 1981): 3–15.

Bernath, Michael T. *Confederate Minds: The Struggle for Intellectual Independence in the Civil War South.* Chapel Hill: University of North Carolina Press, 2010.

Bilgrami, Akeel. "Notes Toward the Definition of 'Identity.'" *Daedalus* 135, no. 4 (Fall 2006): 5–14.

Blight, David W. *Beyond the Battlefield: Race, Memory, and the American Civil War.* Amherst: University of Massachusetts Press, 2002.

Bradley, Mark L. *The Army and Reconstruction, 1865–1877.* Washington, D.C.: U.S. Army Center of Military History, 2015.

Brass, Paul R. *Ethnicity and Nationalism: Theory and Comparison.* London: Sage, 1991.

Brown, Canter Jr. *Ossian Bingley Hart: Florida's Loyalist Reconstruction Governor.* Baton Rouge: Louisiana State University Press, 1997.

Brown, Philip Marshall. "Postwar Reconstruction." *World Affairs* 170, no. 1 (Summer 2007): 33–35.

Brubaker, Rogers. *Ethnicity without Groups.* Cambridge, MA: Harvard University Press, 2004.

Buck, Paul H. *The Road to Reunion 1865–1900.* New York: Vintage Books, 1937.

Budiansky, Stephen. *The Bloody Shirt: Terror After Appomattox.* New York: Viking Penguin, 2008.

Buel, Clarence, and Robert U. Johnson, eds. *The Way to Appomattox: Battles and Leaders of the Civil War Vol IV.* New York: Castle Books, 1956.

Buffaloe, David L. "Defining Asymmetric Warfare." The Land Warfare Papers, no. 58 Arlington, VA: The Institute of Land Warfare, Association of the United States Army, 2006.

Burton, Vernon. "Race and Reconstruction: Edgefield County, South Carolina." *Journal of Southern History* 12, no. 1 (Autumn 1978): 31–56.

Bush, Kenneth D., and E. Fuat Keyman. "Identity-Based Conflict: Rethinking Security in a Post-Cold War World." *Global Governance* 3, no. 3 (September–December 1997): 311–28.

Calhoun, Charles W. *Conceiving a New Republic: The Republican Party and the Southern Question, 1869–1900*. Lawrence: University Press of Kansas, 2006.

Carpenter, Jesse T. *The South as a Conscious Minority, 1789–1861: A Study in Political Thought*. Columbia: University of South Carolina Press, 1930.

Carpenter, John. "Atrocities during the Reconstruction Period." *Journal of Negro History* 47 (1962): 234–47.

Carter, Dan T. *When the War Was Over: The Failure of Self-Reconstruction in the South, 1865–1867*. Baton Rouge: Louisiana State University Press, 1985.

Chamberlain, E. J. R. " Asymmetry: What Is It and What Does It Mean for the British Armed Forces?" *Defence Studies* 3, no. 1 (Spring 2003): 17–43.

Chirwa, Wiseman. "Collective Memory and the Process of Reconciliation and Reconstruction." *Development in Practice* 7, no. 4 (November 1997): 479–82.

Crow, Jeffrey J. "Thomas Settle, Jr., Reconstruction, and the Memory of the Civil War." *Journal of Southern History* 12, no. 4 (November 1996): 689–726.

Cunningham, David E., Kristin Skrede Gleditsch, and Idean Salehyan. "It Takes Two: A Dyadic Analysis of Civil War Duration and Outcome." *Journal of Conflict Resolution* 53, no. 4 (August 2009): 570–97.

Cunningham, Kathleen Gallagher. "Actor Fragmentation and Civil War Bargaining: How Internal Divisions Generate Civil Conflict." *American Journal of Political Science* 57, no. 3 (July 2013): 659–72.

Curry, Richard O., ed. *Radicalism, Racism, and Party Realignment: The Border States During Reconstruction*. Baltimore: Johns Hopkins University Press, 1969.

Dabney, Virginius. *Virginia: The New Dominion*. New York: Doubleday, 1971.

Daniell, Elizabeth Otto. "The Ashburn Murder Case in Georgia Reconstruction, 1868." *Georgia Historical Quarterly* 59, no. 3 (Fall 1975): 296–312.

Darrow, William B. "The Killing of Congressman James Hinds." *Arkansas Historical Quarterly* 72 (Spring 2015): 18–55.

Davis, Thomas C. "Revisiting Group Attachment: Ethnic and National Identity." *Political Psychology* 20, no. 1 (March 1999): 25–47.

Davison, Kenneth E. *The Presidency of Rutherford B. Hayes*. Westport, CT: Greenwood Press, 1972.

Dawson, Joseph G. III. *Army Generals and Reconstruction, 1862–1877*. Baton Rouge: Louisiana State University Press, 1982.

Deaux, Kay, and Tracy McLaughlin-Volpe. "An Organizing Framework for Collective Identity: Articulation and Significance of Multidimensionality." *Psychological Bulletin* 130, no. 1 (2004): 80–114.

DeBlack, Thomas. *With Fire and Sword: Arkansas 1861–1874*. Fayetteville: University of Arkansas Press, 2003.

DeRouen, Karl R., Jr., and David Subek. "The Dynamics of Civil War Duration and Outcome." *Journal of Peace Research* 41, no. 3 (May 2004): 303–20.

Digital Library of Georgia. Civil War Unrest in Camilla, Georgia, 1868. https://dlg.galileo.usg.edu/camilla (accessed July 26, 2016).

Dollar, Kent T., Larry H. Whittaker, and W. Calvin Dickinson, eds. *Sister States, Enemy States: The Civil War in Kentucky and Tennessee.* Lexington: University of Kentucky Press, 2009.

Downs, Gregory P. *After Appomattox: Military Occupation and the Ends of War.* Cambridge, MA: Harvard University Press, 2015.

Downs, Gregory P. *The Declaration of Dependence: The Long Reconstruction of Popular Politics in the South, 1861–1908.* Chapel Hill: University of North Carolina Press, 2011.

Downs, Gregory P., and Kate Masur, eds. *The World the Civil War Made.* Chapel Hill: University of North Carolina Press, 2015.

Dubiel, Helmut. "Cultivated Conflicts." *Political Theory* 26, no. 2 (April 1998): 209–20.

Dunne, J. Paul, María D. C. García-Alonso, Paul Levine, and Ron P. Smith. "Managing Asymmetric Warfare." *Oxford Economic Papers.* New Series. 58, no. 2 (April, 2006): 183–208.

Dunning, Thad. "Fighting and Voting: Violent Conflict and Electoral Politics." *Journal of Conflict Resolution* 55, no. 3 (June 2011): 327–339.

Dunning, William A. "Military Government in the South during Reconstruction." *Political Science Quarterly* 12, no. 3 (September 1897): 381–406.

Durden, Robert F. "The Prostrate State Revisited: James K. Pike and South Carolina Reconstruction." *Journal of Negro History* 19, no. 2 (April 1954): 87–110.

Durrill, Wayne K. "Political Legitimacy and Local Courts: 'Politicks at Such a Rage' in a Southern Community During Reconstruction." *Journal of Southern History* 70, no. 3 (August 2004): 577–602.

Eaton, J. G. "The Beauty of Asymmetry: An Examination of the Context and Practice of Asymmetric and Unconventional Warfare from a Western/Centrist Perspective." *Defence Studies* 2, no. 1 (Spring 2002): 51–82.

Egerton, Douglas R. *The Wars of Reconstruction: The Brief, Violent History of America's Most Progressive Era.* New York: Bloomsbury Press, 2014.

Emberton, Carole. *Beyond Redemption: Race, Violence, and the American South after the Civil War.* Chicago: University of Chicago Press, 2013.

Engle, Stephen D. "Mountaineer Reconstruction: Blacks in the Political Reconstruction of West Virginia." *Journal of Negro History* 78, no. 3 (Summer 1993): 137–165.

Faust, Drew Gilpin. *The Creation of Confederate Nationalism: Ideology and Identity in the Civil War South.* Baton Rouge: Louisiana State University Press, 1988.

Fearon, James D., and David D. Laitin. "Ethnicity, Insurgency, and Civil War." *American Political Science Review* 97, no. 1 (February 2003): 75–90.

Findley, Michael G., and Scott Edwards. "Weak-Actor Social Structure in Asymmetric Wars." *International Studies Quarterly* 51, no. 3 (September, 2007): 583–606.

Fitzgerald, Michael W. "Property Crime and the Plantation System in Reconstruction Alabama." *Agricultural History Society* 71, no. 2 (Spring 1997): 186–206.

Fitzgerald, Michael W. "Radical Republicanism and the White Yeomanry During Alabama Reconstruction, 1865–1868." *Journal of Southern History* 54, no. 4 (November 1988): 565–596.

Fitzgerald, Michael W. *Splendid Failure: Postwar Reconstruction in the American South.* The American Ways Series. Chicago: Ivan R. Dee, 2007.

Fitzgerald, Michael W. *The Union League Movement in the Deep South: Politics and Agricultural Change During Reconstruction.* Baton Rouge: Louisiana State University Press, 2000.

Fletcher, Marvin. "The Negro Volunteer in Reconstruction, 1865–66." *Military Affairs* 32, no. 3 (December 1968): 124–31.

Flores, Thomas Edward, and Irfan Nooruddin. "The Effect of Elections on Postconflict Peace and Reconstruction." *Journal of Politics* 74, no. 2 (April 2012): 558–70.

Fogelson, Robert M., and Richard E. Rubenstein, eds. *Use of the Army in Certain Southern States.* New York: Arno Press, 1969.

Fominaya, Christine Flesher. "Collective Identity in Social Movements: Central Concepts and Debates." *Sociology Compass* 4, issue 6 (2010): 393–404.

Foner, Eric. *Reconstruction: America's Unfinished Revolution, 1863–1877.* New York: Harper and Row, Publishers, 1988.

Foner, Eric. "Reconstruction Revisited." *American History* 10, no. 4 (December 1982): 82–100.

Formwalt, Lee W. "The Camilla Massacre of 1868: Racial Violence as Political Propaganda." *Georgia Historical Quarterly* 71, no. 3 (Fall, 1987): 399–426.

Freeman, Michael, and Hy Rothstein, eds. *Gangs and Guerrillas: Ideas from Counterinsurgency and Counterterrorism.* Monterrey, CA: Naval Postgraduate School (March 2011).

Gaston, Paul M. *The New South Creed: A Study in Southern Mythmaking.* New York: Alfred A. Knopf, 1970.

Gillette William. *Retreat From Reconstruction 1869–1879.* Baton Rouge: Louisiana State University Press, 1982.

Gillin, Kate Côté. *Shrill Hurrahs: Women, Gender, and Racial Violence in South Carolina, 1865–1900.* Columbia: University of South Carolina Press, 2013.

Gray, Colin S. "Thinking Asymmetrically in Times of Terror." *Parameters* (Spring 2002): 5–14.

Gross, Michael L. "Asymmetric War, Symmetrical Intentions: Killing Civilians in Modern Armed Conflict." *Global Crime* 10, no. 4 (November 2000): 320–36.

Hale, Grace Elizabeth. *Making Whiteness: The Culture of Segregation in the South, 1890–1940.* New York: Pantheon Books, 1998.

Harcourt, Edward John. "The Whipping of Richard Moore: Reading Emotion in Reconstruction America." *Journal of Southern History* 36, no. 2 (Winter 2002): 261–82.

Hardy, William Edward, "'Fare Well to All Radicals': Redeeming Tennessee, 1869–1870." PhD diss., University of Tennessee, 2013.

Harris, William C. *With Charity for All: Lincoln and the Restoration of the Union.* Lexington: University Press of Kentucky, 1997.

Harris, William C. *The Day of the Carpetbagger: Republican Reconstruction in Mississippi.* Baton Rouge: Louisiana State University Press, 1979.

Hedetoft, Ulf. "National Identities and Mentalities of War in Three EC Countries." *Journal of Peace Research* 30, no. 3 (August 1993): 281–300.

Henderson, Errol A. "Culture or Contiguity: Ethnic Conflict, the Similarity of States, and the Onset of War, 1820–1989." *Journal of Conflict Resolution* 41, no. 5 (October 1997): 649–68.

Herrmann, Richard K., Pierangelo Isernia, and Paolo Segatti. "Attachment to the Nation and International Relations: Dimensions of Identity and Their Relationship to War and Peace." *Political Psychology* 30, no. 5 (October 2009): 721–54.

Higginbotham, F. Michael. *Ghosts of Jim Crow: Ending Racism in Post-Racial America.* New York: New York University Press, 2013.

Hoddie, Matthew, and Caroline Hartzell. "Signals of Reconciliation: Institution-Building and the Resolution of Civil Wars." *International Studies Review* 7, no. 1 (March 2005): 26–30.

Hodes, Martha. "The Sexualization of Reconstruction Politics: White Women and Black Men in the South after the Civil War." *Journal of the History of Sexuality* 3, no. 3 (January 1993): 402–17.

Hogue, James K. *Uncivil War: Five New Orleans Street Battles and the Rise and Fall of Radical Reconstruction.* Baton Rouge: Louisiana State University Press, 2006.

Holt, Michael F. *By One Vote: The Disputed Presidential Election of 1876*. Lawrence: University Press of Kansas, 2008.

Hoogenboom, Ari. *The Presidency of Rutherford B. Hayes*. Lawrence: University Press of Kansas, 1988.

House of Representatives, State of Mississippi. *The Testimony in the Impeachment of Adelbert Ames as Governor of Mississippi*. Jackson, MS: Power and Barksdale, State Printers, 1877.

Howell, Kenneth W., ed. *Still the Arena of Civil War: Violence and Turmoil in Reconstruction Texas, 1865–1874*. Denton: University of North Texas Press, 2012.

Hume, Richard L. "The Arkansas Constitutional Convention of 1868: A Case Study in the Politics of Reconstruction." *Journal of Southern History* 39, no. 2 (May 1973): 183–206.

Hume, Richard L. "Carpetbaggers in the Reconstruction South: A Group Portrait of Outside Whites in the 'Black and Tan' Constitutional Conventions." *Journal of American History* 64, no. 2 (September 1977): 313–30.

James, Joseph B. "Southern Reaction to the Proposal of the Fourteenth Amendment." *Journal of Southern History* 22, no. 4 (November 1956): 477–97.

Janney, Caroline E. "The Right to Love and Mourn: The Origins of Virginia's Ladies' Memorial Associations, 1865–1867." In *The Crucible of Civil War: Virginia from Secession to Commemoration*, edited by Edward L. Ayers, Gary W. Gallagher, and Andrew L. Torget, pp. 165–188. Charlottesville: University of Virginia Press, 2006.

Kalyvas, Stathis N. *The Logic of Violence in Civil War*. New York: Cambridge University Press, 2006.

Kalyvas, Stathis N. "The Ontology of Political Violence: Action and Identity in Civil Wars." *Perspectives on Politics* 1, no. 3 (September 2003): 475–94.

Kalyvas, Stathis N. "Warfare in Civil Wars." In *Rethinking the Nature of War*, edited by Isabelle Duyvesteyn and Jan Angstrom, pp. 88–108. Abingdon, VA: Frank Cass, 2005.

Kalyvas, Stathis N., and Matthew Adam Kocher. "How 'Free' is Free Riding in Civil Wars? Violence, Insurgency, and the Collective Action Problem." *World Politics* 59, no. 2 (January 2007): 177–216.

Keen, David. *Useful Enemies: When Waging Wars Is More Important Than Winning Them*. New Haven, CT: Yale University Press, 2012.

Kekes, John. "War." *Philosophy* 85, no. 332 (April 2010): 201–18.

Kennedy-Nolle, Sharon D. *Writing Reconstruction: Race, Gender and Citizenship in the Postwar South*. Chapel Hill: University of North Carolina Press, 2015.

King, Ronald F. "Counting the Votes: South Carolina's Stolen Election of 1876." *Journal of Interdisciplinary History* 32, no. 2 (Fall 2001): 169–91.

Kiss, Peter. *Winning Wars amongst the People: Case Studies in Asymmetric Conflict.* Lincoln, NE: Potomac Books, 2014.

Kissane, Bill, ed. *After Civil War: Diversion, Reconstruction and Reconciliation in Contemporary Europe.* Philadelphia: University of Pennsylvania Press, 2014.

Knorr, Klaus, ed. *Power, Strategy, and Security: A World Politics Reader.* Princeton, NJ: Princeton University Press, 1983.

Kolchin, Peter. "Carpetbaggers and Reconstruction: A Quantitative Look at Southern Congressional Politics, 1868–1872." *Journal of Southern History* 45, no. 1 (February 1979): 63–76.

Kousser, J. Morgan "Post-Reconstruction Suffrage Restrictions in Tennessee: A New Look at the V.O. Key Thesis." *Political Science Quarterly* 88, no. 4 (December 1973): 655–83.

Kunovich, Robert M., and Randy Hodson. "Conflict, Religious Identity, and Ethnic Intolerance in Croatia." *Social Forces* 78, no. 2 (December 1999): 643–68.

Lane, Charles. *The Day Freedom Died: The Colfax Massacre, the Supreme Court, and the Betrayal of Reconstruction.* New York: Henry Holt, 2008.

Lehman, Nicholas. *Redemption: The Last Battle of the Civil War.* New York: Farrar, Straus and Giroux, 2006.

Lyall, Jason, and Isiah Wilson, III. "Rage Against the Machine: Explaining Outcomes in Counterinsurgency Wars." *International Organization* 63, no. 1 (Winter 2009): 67–106.

MacGinty, Roger. "The Pre-War Reconstruction of Post-War Iraq." *Third World Quarterly* 24, no. 4 (August 2003): 601–17.

Mack, Andrew. "Why Nations Lose Small Wars: The Politics of Asymmetric Conflict." *World Politics* 27, no. 2 (January 1975): 175–200.

Mack, Raymond W., and Richard C. Snyder. "The Analysis of Social Conflict—Toward an Overview and Synthesis." *Conflict Resolution* 1, no. 2 (June 1957): 212–48.

Martinez, J. Michael. *Carpetbaggers, Cavalry, and the Ku Klux Klan.* New York: Rowman and Littlefield, 2007.

Mason, T. David, Joseph P. Weingarten, Jr., and Patrick J. Felt. "Win, Lose, or Draw: Predicting the Outcome of Civil Wars." *Political Research Quarterly* 78, no. 2 (June 1999): 239–68.

McCardell, John. *The Idea of a Southern Nation: Southern Nationalists and Southern Nationalism, 1830–1860.* New York: W.W. Norton, 1979.

McCormick, Gordon H., and Frank Giordano. "Things Come Together: Symbolic Violence and Guerrilla Mobilisation." *Third World Quarterly* 28, no. 2 (March 2007): 295–320.

Meade, George G. *Major General Meade's Report on the Ashburn Murder.* Atlanta: U.S. Army Department of the South, 1868.

Melucci, Alberto. "The Process of Collective Identity." In *Social Movements and Culture*, edited by Hank Johnson and Bert Klandermans, pp. 41–63. Minneapolis: University of Minnesota Press, 1995.

Mendelson, Wallace. "A Note on the Cause and Cure of the Fourteenth Amendment." *Journal of Politics* 43, no. 1 (February 1981): 152–58.

Miles, Franklin B. "Asymmetric Warfare: An Historical Perspective." USAWC Strategy Research Project, Carlisle, PA: U.S. Army War College, 1999.

Mitchell, C. R. "Evaluating Conflict." *Journal of Peace Research* 17, no. 1 (1980): 61–75.

Moneyhon, Carl. *The Impact of the Civil War and Reconstruction on Arkansas: Persistence in the Midst of Ruin.* Baton Rouge: Louisiana State University Press, 1994.

Morris, Roy, Jr. *Fraud of the Century: Rutherford B. Hayes, Samuel Tilden, and the Stolen Election of 1876.* New York: Simon and Schuster, 2003.

Nash, Steven E. *Reconstruction's Ragged Edge: The Politics of Postwar Life in the Southern Mountains.* Chapel Hill: University of North Carolina Press, 2016.

Nelson, Scott Reynolds. *Iron Confederacies: Southern Railways, Klan Violence and Reconstruction.* Chapel Hill: University of North Carolina Press, 1998.

O'Brien, Michael. *The Idea of the American South, 1920–1941.* Baltimore: Johns Hopkins University Press, 1979.

Olsen, Otto H., ed. *Reconstruction and Redemption in the South.* Baton Rouge: Louisiana State University Press, 1980.

Oser, Jacob. *Henry George.* New York: Twayne Publishers, 1974.

Paasi, Ansi. "Boundaries as Social Processes: Territoriality in the World of Flows." In *Boundaries, Territory and Postmodernity*, edited by David Newman, pp. 69–88. London: Frank Cass, 1999.

Parsons, Elaine Frantz. *Ku-Klux: The Birth of the Klan during Reconstruction.* Chapel Hill: University of North Carolina Press, 2015.

Parsons, Elaine Frantz. "Midnight Rangers: Costume and Performance in the Reconstruction-era Ku Klux Klan." *Journal of American History* 92, no. 3 (December 2005): 811–36.

Perman, Michael. *Reunion without Compromise: The South and Reconstruction: 1865–1868.* CUP Archive, 1973.

Perman, Michael. *The Road to Redemption: Southern Politics, 1869–1879*. Chapel Hill: University of North Carolina Press, 1985.

Perman, Michael. *Struggle for Mastery: Disfranchisement in the South 1888–1908*. Chapel Hill: University of North Carolina Press, 2001.

Pettegrew, John. "The Soldier's Faith: Turn-of-the-Century Memory of the Civil War and the Emergence of Modern American Nationalism." *Journal of Contemporary History*, 31 (1996): 49–73.

Pfaff, Steven. "Collective Identity and Informal Groups in Revolutionary Mobilization: East Germany in 1989." *Social Forces* 75, no. 1 (September 1996): 91–117.

Pierce, Edward. *Memoir and Letters of Charles Sumner*. Vol. 4. Cambridge, MA: Press of John Wilson and Son, 1877.

Polletta, Francesca, and James M. Jasper. "Collective Identity and Social Movements." *Annual Review of Sociology* 27 (2007): 283–305.

Poole, W. Scott. *Never Surrender: Confederate Memory and Conservatism in the South Carolina Upcountry*. Athens: University of Georgia Press, 2004.

President's Committee on Civil Rights. *To Secure These Rights: The Report of the President's Committee on Civil Rights*. Washington, D.C.: U.S. Government Printing Office, 1947.

Prince, K. Stephen. *Stories of the South: Race and the Reconstruction of Southern Identity, 1865–1915*. Chapel Hill: University of North Caroline Press, 2014.

Rable, George C. *But There Was No Peace: The Role of Violence in the Politics of Reconstruction*. Athens: University of Georgia Press, 2007.

Rawley, James A. "The General Amnesty Act of 1872: A Note." *The Mississippi Valley Historical Review* 47, no. 3 (December 1960): 480–84.

Rehnquist, William H. *Centennial Crisis: The Disputed Election of 1876*. New York: Alfred A. Knopf, 2004.

Richardson, Heather Cox. *The Death of Reconstruction: Race, Labor, and Politics in the Post-Civil War North, 1865–1901*. Cambridge, MA: Harvard University Press, 2001.

Richardson, Heather Cox. *West from Appomattox: The Reconstruction of America after the Civil War*. New Haven, CT: Yale University Press, 2007.

Riddleberger, Patrick W. "The Radicals' Abandonment of the Negro during Reconstruction." *Journal of Negro History* 45, no. 2 (April 1960): 88–102.

Robson, Charles. "Col. D. M. Carter North Carolina." In *Representative Men of the South*. Philadelphia: Charles Robson, 1880, 418–20. http://archive.org/stream/representativeme00robs#page/418/mode/2up (accessed June 21, 2016).

Robinson, Armistead L. "The Politics of Reconstruction." *Wilson Quarterly* 2, no. 2 (Spring 1978): 106–23.

Ross, Marc Howard. "A Cross-Cultural Theory of Political Conflict and Violence." *Political Psychology* 7, no. 3 (September 1986): 427–69.

Ross, Marc Howard. "Internal and External Conflict and Violence: Cross-Cultural Evidence and a New Analysis." *Journal of Conflict Resolution* 29, no. 4 (December 1985): 547–79.

Rothman, Jay, and Marie L. Olson. "From Interests to Identities: Towards a New Emphasis in Interactive Conflict Resolution." *Journal of Peace Research* 38, no. 3 (May 2001): 289–305.

Royster, Charles, ed. *Memoirs of General W. T. Sherman*. New York: Viking Press, 1990.

Rubin, Anne Sarah. *A Shattered Nation: The Rise and Fall of the Confederacy, 1861–1868*. Chapel Hill: University of North Carolina Press, 2005.

Russ, William A., Jr. "The Negro and White Disfranchisement during Radical Reconstruction." *Journal of Negro History* 19, no. 2 (April 1934): 171–92.

Russ, William A., Jr. "Registration and Disfranchisement Under Radical Reconstruction." *Mississippi Valley Historical Review* 21, no. 2 (September 1934): 163–80.

Russ, William A., Jr. "Was There Danger of a Second Civil War During Reconstruction?" *Mississippi Valley Historical Review* 25, no. 1 (June 1938): 39–58.

Schroefl, Josef, and Stuart Kaufman. "Hybrid Actors, Tactical Variety: Rethinking Asymmetric and Hybrid War." *Studies in Conflict and Terrorism* 37, no. 10 (2014): 862–80.

Scott, Thomas A., ed. *Cornerstones of Georgia History: Documents the Formed the State*. Athens: University of Georgia Press, 1995.

Severance, Ben H. *Tennessee's Radical Army: The State Guard and Its Role in Reconstruction, 1867–1869*. Knoxville: University of Tennessee Press, 2005.

Shapiro, Herbert. "The Ku Klux Klan during Reconstruction: the South Carolina Episode." *Journal of Negro History* 49, no. 1 (January 1964): 34–55.

Shofner, Jerrell H. *Nor Is It Over Yet: Florida in the Era of Reconstruction, 1863–1877*. Gainesville: University Press of Florida, 1974.

Siani-Davies, Peter, and Stefanos Katsikas. "National Reconciliation after Civil War: The Case of Greece." *Journal of Peace Research* 46, no. 4 (July 2009): 559–75.

Simkins, Francis B. "The Ku Klux Klan in South Carolina, 1868–1871." *Journal of Negro History* 12, no. 4 (October 1927):606–47.

Simpson, Graeme. "Reconstruction and Reconciliation: Emerging from Transition." *Development in Practice* 7, no. 4 (November 1997): 475–78.

Singletary, Otis. *Negro Militia and Reconstruction.* New York: McGraw Hill, 1957.

Slap, Andrew L. *The Doom of Reconstruction: Liberal Republicans in the Civil War Era.* New York: Fordham University Press, 2006.

Smith, Alastair, and Allan C. Stam. "Bargaining and the Nature of War." *Journal of Conflict Resolution* 48, no. 6 (December 2004): 783–813.

Smith, George P. "Republican Reconstruction and Section Two of the Fourteenth Amendment." *Western Political Quarterly* 23, no. 4 (December 1970): 829–53.

Smith, W. Calvin. "The Reconstruction 'Triumph' of Rufus B. Bullock." *Georgia Historical Quarterly* 52, no. 4 (December, 1968): 414–25.

Snow, David A., and Robert D. Benford. "Ideology, Frame Resonance, and Participant Mobilization." *International Social Movement Research* 1 (1988):197–218.

Spinner-Halev, Jeff, and Elizabeth Theiss-Morse. "National Identity and Self-Esteem." *Perspective on Politics* 1, no. 3 (September 2003): 515–32.

Sproat, John G. "Blueprint for Radical Reconstruction." *Journal of Southern History* 23, no. 1 (February 1957): 25–44.

Stampp, Kenneth M. *The Era of Reconstruction 1865–1877.* New York: Vintage Books, 1965.

Stanton, Amanda. The Encyclopedia of Arkansas History & Culture s. v. James Hinds. http://www.encyclopediaofarkansas.net (accessed September 4, 2016).

Stepanova, Ekaterina. *Terrorism in Asymmetrical Conflict: Ideological and Structural Aspects.* SIPRI Research report no. 23. New York: Oxford University Press, 2008.

Stets, Jan E., and Peter J. Burke. "Identity Theory and Social Identity Theory." *Social Psychology Quarterly* 63, no. 3 (September 2000): 224–37.

Sullivan, Patricia L. "War Aims and War Outcomes: Why Powerful States Lose Limited Wars." *Journal of Conflict Resolution* 51, no. 3 (June 2007): 496–524.

Summers, Mark Wahlgren. *The Ordeal of the Reunion: A New History of Reconstruction.* Littlefield History of the Civil War Era Series. Chapel Hill: University of North Carolina Press, 2014.

Swinney, Everette. "Enforcing the Fifteenth Amendment." *Journal of Southern History* 28, no. 2 (May 1962): 202–18.

Taylor, A. A. "Opposition to the Reconstruction." *Journal of Negro History* 9, no. 4 (October 1924): 442–68.

Taylor, Joe Gray. *Louisiana Reconstructed, 1863–1877.* Baton Rouge: Louisiana State University Press, 1974.

Thomas, Emory M. *The Confederacy as a Revolutionary Experience.* Columbia: University of South Carolina Press, 1971.

Trelease, Alan W. *White Terror: The Ku Klux Klan Conspiracy and Southern Reconstruction.* Baton Rouge: Louisiana State University Press, 1999.

Tunnell, Ted. "Creating 'the Propaganda of History': Southern Editors and the Origins of *Carpetbagger* and *Scalawag.*" *Journal of Southern History* 72, no. 4 (November 2006): 789–822.

Tunnell, Ted. *Crucible of Reconstruction: War, Radicalism and Race in Louisiana 1862–1877.* Baton Rouge: Louisiana State University Press, 1984.

U.S. Congress. House. *Message of the President of the United States and Accompanying Documents to the Two Houses of Congress,* Report of the Secretary of War. 40th Cong., 3d sess. Washington, D.C.: 1868.

U.S. Congress. *Report of the Joint Committee on Reconstruction.* 39th Cong., 1st sess. Washington, D.C.: 1866.

U.S. Congress. *Report of the Joint Select Committee to Inquire into the Condition of Affairs in the Late Insurrectionary States.* 13 vols. 42d Cong., 2d sess. Washington, D.C.: 1872.

U.S. President. *Messages from the President of the United States to the Two Houses of Congress at the Commencement of the Third Session of the Fortieth Congress with the Reports of the Heads of the Departments and Selections from Accompanying Documents.* Washington, D.C.: 1869.

U.S. War Department. *The War of Rebellion: A Compilation of the Official Records of the Union and Confederate Armies.* Series 1, Vol. 46, part 3. Washington, D.C.: U.S. Government Printing Office, 1895.

Varshney, Ashutosh. "Nationalism, Ethnic Conflict, and Rationality." *Perspectives on Politics* 1, no. 1 (March 2003): 85–99.

Vral, Susannah J., ed. *Civil War Citizens: Race, Ethnicity, and Identity in America's Bloodiest Conflict.* New York: New York University Press, 2016.

Vuckovic, Gojko. "Promoting Peace and Democracy in the Aftermath of the Balkan Wars: Comparative Assessment of the Democratization and Institution-Building Processes in Croatia, Bosnia and Herzegovina and Former Yugoslavia." *World Affairs* 162, no. 1 (Summer 1999): 3–10.

Wade, Wyn Craig. *The Fiery Cross: The Ku Klux Klan in America.* New York: Oxford University Press, 1987.

Weinfeld, Daniel R. *The Jackson County War: Reconstruction and Resistance in Post-Civil War Florida.* Tuscaloosa: University of Alabama Press, 2012.

Wendt, Alexander. "Collective Identity Formation and the International State." *The American Political Science Review* 88, no. 2 (June 1994): 384–96.

West, Jerry L. *The Bloody South Carolina Election of 1876: Wade Hampton III, the Red Shirt Campaign for Governor and the End of Reconstruction.* Jefferson, NC: McFarland, 2011.

Wetherington, Mark V. *Plain Folk's Fight: The Civil War and Reconstruction in Piney Woods Georgia.* Chapel Hill: University of North Carolina Press, 2005.

Wilbur, W. Allan. "Joint Committee on Reconstruction, 1865." In *Congress Investigates: A Documented History, 1792–1974*, edited by Arthur M. Schlesinger Jr. and Roger Bruns, pp. 1361–1378. New York: Chelsea House, 1975.

Williams, Andrew J. "Reconstruction before the Marshall Plan." *Review of International Studies* 31, no. 3 (July 2005): 541–58.

Williams, Lou Falkner. *The Great South Carolina Ku Klux Klan Trials, 1871–1872.* Athens: University of Georgia Press, 1996.

Woodward, C. Vann. *The Origins of the New South.* Baton Rouge: Louisiana State University Press, 1951.

Wyatt-Brown, Bertram. *The Shaping of Southern Culture: Honor, Grace, and War, 1760s–1880s.* Chapel Hill: University of North Carolina Press, 2001.

Zipf, Karin L. "'The Whites Shall Rule the Land or Die': Gender, Race, and Class in North Carolina." *Journal of Southern History* 65, no. 3 (August 1999): 499–534.

NEWSPAPERS

California:

Daily Alta California, San Francisco, California

Los Angeles Daily Herald, Los Angeles, California

Delaware:

The Daily Gazette, Wilmington, Delaware

Illinois:

The Cairo Bulletin, Cairo, Illinois

The Cairo Evening Bulletin, Cairo, Illinois

The Daily Argus, Rock Island, Illinois

The Evansville Journal, Evansville, Illinois

Indiana:

The Indiana State Sentinel, Indianapolis, Indiana

Jasper Weekly Courier, Jasper, Indiana

Kansas:

The Leavenworth Weekly Times, Leavenworth, Kansas

The Smoky Hill and Republican Union, Junction City, Kansas

Kentucky:

The Hartford Herald, Hartford, Kentucky

Louisiana:

The Bossier Banner, Bellevue, Louisiana

The Donaldsville Chief, Donaldsville, Louisiana

The Louisiana Democrat, Alexandria Louisiana

The Louisiana Democrat, New Orleans, Louisiana

The New Orleans Bulletin, New Orleans, Louisiana

The New Orleans Crescent, New Orleans, Louisiana

The New Orleans Daily Democrat, New Orleans, Louisiana

New Orleans Republican, New Orleans, Louisiana

The Opelousas Courier, St. Landry Parish, Louisiana

The Shreveport Times, Shreveport, Louisiana

The South-Western, Shreveport, Louisiana

The Weekly Echo, Lake Charles Parish, Calcasieu, Louisiana

Maryland:

The Aegis and Intelligencer, Bel Air, Maryland

The Cecil Whig, Elkton, Maryland

The Democratic Advocate, Westminster, Maryland

Massachusetts:

The Boston Daily Globe, Boston, Massachusetts

Michigan:

The True Northerner, Paw Paw, Michigan

Minnesota:

St. Paul Daily Globe, St. Paul Minnesota

Mississippi:

The Daily Clarion, Jackson, Mississippi

Jackson Pilot, Jackson, Mississippi

The Weekly Clarion, Jackson, Mississippi

Montana:
Helena Weekly Herald, Helena, Montana

Nevada:
The Daily State Register, Carson City, Nevada

New York:
The New York Herald, New York City, New York
The New York Times, New York City, New York
New York Tribune, New York City, New York
The New York World, New York City, New York

North Carolina:
Carolina Watchman, Salisbury, North Carolina
The News and Herald, Winnsboro, North Carolina
The Old North State, Salisbury, North Carolina
The Southerner, Tarboro, North Carolina
The Tri-Weekly Standard, Raleigh, North Carolina
Weekly North-Carolina Standard, Raleigh, North Carolina
The Weekly Standard, Raleigh, North Carolina
The Western Democrat, Charlotte, North Carolina
Western Sentinel, Winston-Salem, North Carolina
Wilmington Journal, Wilmington, North Carolina

Ohio:
Ashtabula Telegraph, Ashtabula, Ohio
Commercial, Cincinnati, Ohio
Delaware Gazette, Delaware, Ohio
The Democratic Press, Ravenna, Ohio
Fayette County Herald, Washington Court House, Ohio
Gallipolis Journal, Gallipolis, Ohio
The Highland Weekly News, Hillsboro, Ohio
Jackson Standard, Jackson Court House, Ohio
Stark County Democrat, Canton, Ohio
The States and Union, Ashland, Ohio
Western Reserve Chronicle, Warren, Ohio

Pennsylvania:
American Citizen, Butler, Pennsylvania
The Centre Reporter, Centre, Pennsylvania
Clearfield Republican, Clearfield, Pennsylvania
The Columbia Freeman, Ebensburg, Pennsylvania
The Daily Evening Telegraph, Philadelphia, Pennsylvania
Democrat and Sentinel, Ellensburg, Pennsylvania
The Evening Telegraph, Philadelphia, Pennsylvania
Forest Republican, Tionesta, Pennsylvania
The Jeffersonian, Stroudsburg, Pennsylvania
The Somerset Herald, Somerset, Pennsylvania
Sunbury American, Sunbury, Pennsylvania
Wyoming Democrat, Tunkhannock, Pennsylvania

South Carolina:
The Abbeville Press and Banner, Abbeville, South Carolina
Anderson Intelligencer, Anderson Court House, South Carolina
Charleston Daily News, Charleston, South Carolina
The Daily Phoenix, Columbia, South Carolina
The Edgefield Advertiser, Edgefield, South Carolina
The Fairfield Herald, Winnsboro, South Carolina
Keowee Courier, Pickens Court House, South Carolina
The New South, Beaufort, South Carolina
The Newberry Herald, Newberry, South Carolina
Yorkville Courier, Yorkville, South Carolina
The Yorkville Enquirer, Yorkville, South Carolina

Tennessee:
The Athens Post, Athens, Tennessee
The Bolivar Bulletin, Bolivar, Tennessee
Knoxville Weekly Chronicle, Knoxville, Tennessee
Knoxville Whig, Knoxville Tennessee
The Memphis Daily Appeal, Memphis, Tennessee
The Morristown Gazette, Morristown, Tennessee
Nashville Union and American, Nashville, Tennessee

Nashville Union and Dispatch, Nashville, Tennessee
Public Ledger, Memphis, Tennessee
The Pulaski Citizen, Pulaski, Tennessee
Union Flag, Jonesboro, Tennessee

Texas:
Dallas Herald, Dallas, Texas
Weekly Democratic Statesman, Austin, Texas

Virginia:
Alexandria Gazette, Alexandria, Virginia
The Daily Dispatch, Richmond, Virginia
Richmond Dispatch, Richmond, Virginia
Richmond Times, Richmond, Virginia
Staunton Spectator and General Advertiser, Staunton, Virginia

Vermont:
Lamoille Newsdealer, Hyde Park, Vermont

Washington, D.C.:
The National Era, Washington, D. C.
National Republican, Washington, D.C.
New National Era, Washington, D.C.

West Virginia:
The Weekly Register, Point Pleasant, West Virginia
The Wheeling Intelligencer, Wheeling West Virginia

Index

Abbeville County, South Carolina, 63
Akerman, Amos T., 87, 90
Alabama: congressional hearings in,
   85; disfranchisement, 35; elections
   in, 51; Fifteenth Amendment, 70;
   Ku Klux in, 81; politics in, 90, 91;
   resistance in, 124, 135; readmission,
   49; Reconstruction in, 70; returned
   to white Southern control, 119; state
   constitution in, 51; state militia in
   the New South, 145; unionism in,
   32; violence in, 85
Alamance County, North Carolina, 95,
   96
Ames, Adelbert, 106, 107, 108, 109,
   110, 111
Amnesty Act (1872), 90, 91
Amnesty: policy (Abraham Lincoln),
   14; (Andrew Johnson), 17, 36; Liberal
   Republicans and, 122
Appomattox, 10, 38, 42, 142, 149
Archer, Stevenson, 113
Arkansas: carpetbaggers, 123;
   constitutional convention in, 51;
   disfranchisement, 76, 90; elections
   in, 17, 51; Fifteenth Amendment, 70;
   Ku Klux in, 71, 75, 81; martial law in;
   military occupation, 13; politics in,
   51, 74, 75, 99; Reconstruction in, 70;
   resistance in, 60, 73, 112; returned
   to white Southern control, 119; state

militia in, 71, 73, 74, 77, 79, 145;
   unionism in, 32; violence in, 32, 63;
   wartime occupation in, 107
Ashburn, George W., 63
Asymmetric warfare: collective
   identity and, xviii, 7, 9, 12, 37, 38, 40,
   41, 44, 64, 71, 76, 85, 93, 106, 116,
   141, 154; conditions, xviii, 12, 38,
   149; conduct of, 1, 2, 3, 4, 44, 127,
   131; definition, 2; different from
   insurgency and guerrilla warfare, 7;
   dominant (stronger) actor in, xvii, 1,
   2, 3, 12, 41, 86, 93, 115, 135, 137, 141,
   149, 153; information battlefield and,
   41, 44, 59, 60, 66, 93, 106, 123, 135,
   154; layered violence within, 66, 98,
   153, 154; mobilization of identity
   and, xvi, xix, 2, 5, 7, 40, 41, 44, 46,
   79, 85, 103, 106, 152; newspapers
   and, xix; nonviolence in, xvii, 12,
   79, 149; phases of, xviii, 4, 5, 6, 43,
   44, 45, 55, 59, 61, 64, 68, 91, 93, 94,
   100, 103, 109, 116, 138, 154, 155;
   precursors, 3, 27, 37, 151, 152;
   transitions within, xix, 11, 12, 36,
   55, 59, 68, 91, 93, 94, 114; purpose
   of, 3, 4, 106; surprise, 2, 3, 12, 38,
   60, 65, 68, 79, 81, 154; political
   equilibrium as a goal of, xvii, 6, 11,
   12; violence in, 3, 5, 8, 12, 44, 45, 46,
   61, 64, 79, 95, 119, 141, 149, 153, 154;

About the Author

Keith D. Dickson is a professor of military studies at the Joint Forces Staff College, National Defense University, teaching military history and operational planning. Dickson served 30 years as an army officer, almost entirely in special operations. He deployed to Iraq, Afghanistan, and the Horn of Africa as the command historian for U.S. Special Operations Command (SOCOM). Dr. Dickson has received awards for distinguished teaching from Virginia Military Institute and Joint Forces Staff College. He is the author of the award-winning book, *Sustaining Southern Identity: Douglas Southall Freeman and Memory in the Modern South*.